C0-AKW-274

DECOLONIZING HISTORY
Technology and Culture in India, China and the West 1492 to the Present Day

DECOLONIZING HISTORY

Technology and Culture in India, China and the West
1492 to the Present Day

CLAUDE ALVARES

The Apex Press, New York, USA
The Other India Press, Goa, India

First published as *Homo Faber: Technology and Culture in India, China and the West: 1500 to the Present Day* in 1979 by Allied Publishers, New Delhi; in 1980, by Martinus Nijhoff, the Hague and Boston.

This 1991 edition published by The Other India Bookstore, Goa, India and The Apex Press, New York, USA.

Library of Congress Cataloging in Publication Data

Alvares, Claude Alphonso.
Homo Faber: technology and culture in India, China and the West from 1500 to the present day.

Bibliography: p.
Includes index.
1. Technology and Civilization. I Title.
CB478.A48 303.4 79-27571

Cover: Pritish Nandy
Design: Arun Sawant
Painting: Baiju Parthan
About the painting: "*The Chinese box and the Lotus combined to express the cultural and technical achievements of the two civilizations*".

ISBN 0-945257-40-6

Printed by S.J. Patwardhan at Mudra, 383 Narayan, Pune 411 030, India *for Eulalia and for Norma*

Contents

Foreword

The relevance of this book goes beyond its scholastic title. It sheds light on and sets in perspective the large questions of our time, questions of both theory and human choice. It is an attempt to relate, "in ever widening spirals of thought", the basic anthropological concern regarding the nature of man and the predicament that faces him, the role of technology in defining this in our time, the dominant cultural paradigm underlying such a relationship between technology and human destiny, and the political processes through which this relationship and its transmission of a particular culture are sought to be legitimized and challenged in our time. It is an ambitious undertaking gently (though not modestly) carried out, a political battle that is intellectually waged.

Keenly aware that the basic paradigmatic issues facing human intellect have been transformed into problems of power and confrontation of cultures, the author seeks to understand the deep chasm that divides societies through examining the intellectual and historical roots thereof. Most of the current debates on technology, development and international order are reflected in the analysis presented here. The ecological crisis and the role of Western technology in it is spelt out vigorously at various points in the analysis. The theme that it is not merely technology that is at fault but the meaning and direction given to it by the cultural paradigm of the modern West is affirmed and in parts demonstrated. The author is aware of the central importance of the politics of technological choices and the international and global structuring thereof; he is at the same time unsparing of the elites and intelligentsia of the Third World for their falling prey to such choices and in the process ravaging their lands and exploiting their peoples.

The author is also aware of the emerging shifts in the world balance of forces and the likelihood of reversals in dependencies and relative advantages between the North and the South. And having exposed the deleterious consequences of the Western model of development, he presses home not only the desirability of alternative development strategies but also their feasibility and necessity. But he is deeply troubled by the persisting halo of the dominant model based on a technology of abundance in large parts of the world, a model that has proved both unworkable in practice and invalid in theory and whose appeal rests almost wholly on illusions and the sway of vested interests.

The picture of reality that emerges from this study and the volume of empirical evidence it draws on is one of a balance of hope and despair – of the possibility, even historical necessity, of transformation of both of the techno-economic structure of the world and its cultural underpinning but one that is constrained by the staying power of the *status quo* and its capacity to gain new adherents and new legitimacy. It points to the critical importance of political consciousness and will – not just of elites but of a whole spectrum of actors and movements – for intervening in this scenario and bringing what seems to be on the cards to fruition. But while it points to the criticality of the political dimension for effective intervention, it does not attempt to provide a strategy for bringing this about. There are many insights and suggestive leads in this direction but these do not add to a strategy of transition that is not just intellectually perspicacious but also politically feasible.

Such a strategy is hard to come by. The available doctrines of social change have become dated. Indeed it is not clear any more that any one compelling, all-encompassing intellectual structure of thought and action can alter the human condition. Nor is it clear that what appear as historically necessary tendencies can in fact prove inevitable – except perhaps 'ultimately'. Human systems are far too complex for a given framework of thought to comprehend and control. New variables are always in the offing, new defences for prevailing systems contending with new crises threatening them.

The best example of this is the fate of Marxism in contemporary history – without doubt the most influential of all intellectual movements. Marx's projected transformation from world capitalism to world socialism is nowhere in sight. It is not that Marx's conception of historical change lacked in either theoretical rigour or empirical sensitivity, or the capacity to spur people to action. His was one system of thought that could claim to have done all three. It was rather that new phenomena shaped the historical process which Marx could not foresee: the long lease of life provided to capitalism by imperialisms of various shades, the success of liberal democratic systems in taming proletariat and other revolutionary pressures in many parts of the world (including in some 'backward' societies), the convergence of opposing systems of thought and belief caused by technological diffusion and the war system to which the very success of communism in Soviet Union and China have contributed. Still more unforeseen defences of the 'system' seem to be on the horizon.

One line of thought in this work, though less fully developed than others, is to show the common heritage of liberalism and Marxism in the techno-economic assumptions of the paradigm of the modern West and to stress the need to look for alternatives outside this cultural paradigm. Marx, like the philosophers he ridiculed, seems better equipped to 'explain' reality than to 'change' it. This book points to the necessity of rejecting the Western pretense of universalism and for non-Western cultures to seek answers to their problems *within*; and in the process not only provide pluralism in techno-cultural systems but, through such pluralism, help Westerners themselves in dealing with the new crop of problems they now encounter. This is a perspective that is beginning to be widely shared. Among other things, it can enable man to transcend the extreme parochialism of Western science and its so-called objectivity, a myth that the author explodes quite ably. Such a transition from the global spread of one particularist culture and hence of one technological paradigm to a diversity of cultural and technological systems will not, however, be politically easy to bring about even if it is intellectually and morally appealing. And the author knows this.

That brings me to what I think is the most important intellectual and political issue arising out of the encounter of cultures that has been posed by this work and the stream of thought it represents. It is the issue of 'relativization' – of technology, of traditions of science, of intellectual paradigms, of culture. The author has forcefully argued and elaborately documented the case that different cultures have historically created their own science, generated techniques relevant to the problems encountered by them in light of the goals accepted by them, and determined these by reference to their distinctive paradigms and meaning systems. He has argued that this continued to be the case until "the disruptive impact of the Vasco da Gama epoch" which traumatized civilizations other than the West, and that it is necessary to re-establish the plurality of paradigms and, thenceforth, of the relationship between technology and culture. There is a great deal of truth in this. The point that with the gradual restoration of independence to various non-Western cultures through their own striving, and the consequent 'relativization' of cultural paradigms and technological selectivity, there will open up enormous possibilities for the peoples of the world to deal with their problems *as perceived by them* has been well argued by the author.

I would like to carry this analysis a step fruther, however. While affirming the value of diversity in realizing human capacities and in preserving the richness and continuity of different cultural streams, it is necessary to also consider the issue at another plane and in the specific historical context of the contemporary world. The expansionism of the West, while lacking in a truly universalistic impulse and based in the main on domination by one local region over the rest of the world, *has* in the process created a 'world history' even if this is not reflected in the writing of modern history, a charge made by Geoffrey Barraclough and examined in some depth by our author here. It is not possible to wish away the West; the point is to expose the illusion perpetrated by it that the Western scientific paradigm provides a universal and hence absolute referent and is the only basis for world unity. It follows that the search for an alternative paradigm has to be a search for a new basis of unity, not merely the assertion of a diversity of cultures and their corresponding scientific and technological traditions.

The material assembled in this volume not only points to a plurality of paradigms encompassing technology and culture and establishes the case for alternative technological systems for the future. It also provides a new comprehension of the human enterprise as a whole, one that rejects the hegemonistic conception of unity and affirms diversity as an essential principle of any attempt at integrating reality. The truth about diversity is better conceived in this manner than as a basis for asserting unique identities and worldviews. The latter approach is not necessarily indefensible. But it provides an insufficient basis for morality. Properly understood, works like this one are attempts to provide a new conceptual comprehension of the human endeavour, oriented to the "exploding perspective of our times" as stated by the author in his Prelude. If this be so, it is necessary to say that cultural relativism is not a satisfactory alternative to hegemonistic universalism.

I shall end by restating my own position on this as I think it is relevant. (I had first stated it in my Editorial introduction to the first issue of *Alternatives : A Journal of World Policy.*) I believe that every major new human experience calls for a new theoretical response, different from all earlier ones: a new theoretical paradigm is not just a mental construct but a response to a new empirical reality. But while this is so – and all assertions of universality for one particular model of life must be resisted on grounds of their being both arrogant and obscurantist – it is also the case that there is some unity underlying diversities.

This is so in two respects, especially in the modern age. First, the different societies one talks of are interrelated and the causes underlying the problems they face are at least in part to be found in these interrelationships. Second, there is a unity in the flow of history which gives it some determinacy, no doubt with a lot of scope for intervention, but not without limits. To ask for a different alternative for each society without reference to the basic unity provided by both the interrelations in which different units stand and the determinacies of the historical process would be to ask for infinite fragmentation which will not stop at national frontiers and will extend to regional, ethnic and even professional identities. Any search for alternatives that does not seek to foster a new and better unity for all is liable to lead to disintegration and chaos. I think it is necessary to reject chaos as a preferred alternative for civil society.

The philosophical perspective that should guide such an endeavour should steer clear of both imperialist claims to universality and the normless striving for relativity: it should affirm both the principle of *autonomy* of each entity (human as well as social) to seek out its own path to self-realization and the principle of *integration of all such* entities in a common framework of interrelationships based on agreed values. The world as it is at present constituted violate both autonomy and integration as principles informing human arrangements. It is a world based, instead, on a framework of dominance that works through endless fragmentation and tensions, one that relies heavily on instruments of violence and institutions of inequity. It is only by seeking a new structure of the world, and as a basis thereof, a new approach to technology and a new ethic of interrelation between cultures and polities that there will be scope for drawing upon the richness of diverse human experiences. and realizing alternative modes of human fulfilment.

RAJNI KOTHARI

Centre for the Study of
Developing Societies

An Apologia as Prelude

> The rising generation will inevitably look back over the twentieth century with different priorities from ours. Born into a world in which – as all present indications suggest – the major questions will not be European questions but the relationships between Europe, including Russia, and America and the peoples of Asia and Africa, they will find little relevance in many of the topics which engrossed the attention of the last generation. The study of contemporary history requires new perspectives and a new scale of values.
>
> – Geoffrey Barraclough

Our descriptive and evaluatory ideas of the various technological systems that men and women have created in different regions of the world and of the cultures involved with them have been formulated, over the past couple of centuries, with reference to the Western experience of these phenomena. Concepts and categories, reflected from a limited area of the human world, have been indiscriminately and illegitimately used to explain, assess, and move all other great chains of being.

While some Western scholars may claim that such actions were set forth in honour, it is a fact that the extension of the personality of the West to the non-Western world can be directly correlated with an increase in the sum total of poverty, pain, and destruction in that part of the globe. We are gradually reaching the stage when it will be possible to proclaim the arrival of a new principle: "the greatest unhappiness of the greatest possible number."

The non-Western experience of the West through the past two centuries and up to the present day, normally subsumed under the mantle of colonial and neo-colonial history, occupies a considerable portion of this book. But I have not intended to repeat what others have already said. On the contrary, I would agree with those historians who tend to see the colonial and neo-colonial disruptions of Southern nations as having been very often exaggerated.

My point is different: that the colonial impact has been small precisely because we are at the beginnings of the real colonial age. The comprehensive disintegration of non-Western societies is yet to come. That traumatic event might only be precluded, in my opinion, if what I term the "Western paradigm" is checked immediately in its influence over the southern real world and its mind: in other words, if it is relativized, or relegated to where it once had its origins.

The unorthodox character of the ideas proposed then in this book is not an inherent quality of the ideas themselves, but owes rather to the unorthodox nature of the framework that has been formulated to contain them. The framework enables, I hope, a sharp criticism of Western images of the world, while simultaneously pushing forward a more positive and objective alternative. The framework itself may appear as being wholly new (and partly audacious), but on no account is it absolutely original. A similar perspective has been outlined in an earlier work, albeit in an area of scholarly endeavour other than mine.

I have here in mind Geoffrey Barraclough's *An Introduction to Contemporary History,*[1] and I am certain it would be in the reader's interest if I did present now, quite briefly, Barraclough's alluring perspective of his field of involvement. For, although our areas of enquiry are not identical (mine is more inclusive, more unwieldy), our points of reference converge, and Barraclough understood, could (but I am not too sure) make my own central concerns more digestible.

In his book, Barraclough sets out to argue a case for a new discipline of study: contemporary history, to be understood as a new phase of historical investigation that takes for its canvas a wider field than the limited Europo-centric preoccupations of modern or conventional history.[2] Contemporary history, he observes, distances itself from its predecessor through the simple discovery that it is *world* history, "and that the forces shaping it cannot be understood unless we are prepared to adopt world-wide perspectives".

The excuse for Barraclough's proposal is his charge, unanswerable, that modern history, in dwelling too exclusively on Europe, has expended more energy than was warranted on the old world that was dying, rather than on the new world that was coming to life. Let me present a random example of what he would criticize: a 1975 edition of a history text, prepared by four scholars, titled, *Civilizations: Western and World.*[3] Of the thirty chapters in the volume, *one* is on Asia! It is difficult to deny that the histories of the past couple of decades have been parochial "accounts of the two world wars, the peace settlement of 1919, the rise of Fascism and National Socialism, and, since 1945, the conflict of the communist and capitalist worlds".

The nature of Barraclough's proposal should not be underrated. He is not merely arguing for more extensive accounts of events, on a world-wide scale: far less does he mean the hasty, embarrassed addition of a few chapters on extra-European affairs. He is demanding, in fact, a new framework and new terms of reference, all of which implies a re-examination and revision of the whole structure of assumptions on which modern history is based. "Precisely because American, African, Chinese, Indian and other branches of extra-European history cut into the past at a different angle, they cut across the traditional lines: and this very fact casts doubt on the adequacy of the old patterns and suggests the need for a new ground-plan."

Finally, and this is where our interests coincide, Barraclough suggests that such a new ground-plan, concerning nothing less than the globalization of the focus of historians, demands the invention of a new scale of values.

Formerly, a modern historian had one central referent: Europe. What happened within Europe was significant; extra-European affairs were literally *extra* – not integral to his concern. Today, European dominance is no longer even theoretically defensible. And any historian, consciously or otherwise setting out to chart the general lines of history, and at the same time fundamentally convinced that the rise of Asia, Africa, Latin America, and the Communist economies can still be treated as tangentially interesting, will severely test the credibility of his readers.

Yet, such an historian *is* faced with the choice of criteria and selectivity, above all else, of reference. In place of one point of reference, he now has half a dozen, even more.[4] Perhaps we have not so much a problem here as we have an almost impossible requirement of absolute impartiality and neutrality, which would place any consequent undertaking beyond the terms of most cultures. Barraclough has succeeded very well here, by abdicating the European perspective. For that reason alone, he might well prove to be the first universal historian of the twentieth century, which means that universities in any part of the world, north and south, could prescribe him with little unease.

There is, however, one crucial difference between Barraclough's intentions and mine, which might well serve to underline what this book is *not* about. Barraclough intends to follow up his ground-plan with an actual narrative history of the contemporary world-picture based on it. I am interested in the mere establishment of a similar ground-plan and the problems this involves, in our understanding of technology and culture. The writing of an actual history of technology, for example, is beyond my intentions, and more important, beyond my competence.

This is not to deny that I have used a considerable quantity of material from the history of technology: but this material should be seen as illuminating a point or the more general themes of the book. The reader should *not* look for any systematic, even representative treatment of the history of technology. Similar observations may be made of the other disciplines encountered in the volume: history, anthropology, sociology, or political science. While discussing the historical development of societies, for example, I have left out detailed considerations of Japan, the United States of America, or the Soviet bloc. Thus, while the book seems ambitious in aim, it is not so in actual fact.

For, unlike Barraclough, I am not an historian. Neither, for that matter, am I an anthropologist. I was trained principally in philosophy, within the "generalist" tradition. I am therefore naturally disposed to enquire into facts, or rather, into the manner in which our values or pre-suppositions determine our approach to and selection of facts. It is a happy coincidence that only "generalists" (an ever dwindling species) dare take on studies of this kind, for I remember Geertz once saying that an accurate adequate understanding of new countries demands that one pursue scientific quarry across any fenced-off academic field into which it may happen to wander.

Further, a fresh understanding of technology and culture and the invention of a new framework for the purpose are only really justified if I can show that existing frameworks have proved inadequate: at numerous places in the book, I shall indicate the inadequacies. Since most of these limitations are easily betrayed in print, I have focused my sights on the *literature* traditionally dealing with the twin issues: therefore, the bookish quality of some of the ideas presented here.

Barraclough is useful here again: he accepts the fact of Europocentric studies as a given, then works to avoid it. I feel it necessary to delve deeper to discover some answer to how we arrived at a situation where Europe could appropriate to itself a position of absolute referent.[5] I believe that *world* history began earlier, about 1800 as an average point, when the balance between Asia and Europe turned into

European dominance over Asia, thereby completing the total rule over the south by the northern hemisphere. I am also keen to examine the lateral "spread" of this idea of dominance from politics to our ideas of technology and culture. In other words, how did Europeans come to believe (unlike the Chinese) that evolution had created two different human minds – one for themselves, another for everybody else?

Here we come across a very crucial problem, contained in the principle enunciated to me (in jest) by Dr. Ward Morehouse, that all cultures may be equal, but some continue to be more equal than others. A culture is a system of values, or as Geertz put it, a web of meanings. By definition, values, meanings and the systems based on or incorporating them are incomparable. There is in reality no independent standard in relation to which one could compare Indian, Chinese, Euro-American, or African culture to come up with a result of one being better or superior to the rest. Let me explain this with an example.

Random example. In the assemblage, maintenance or overhaul of a machine tool, one of the considerations the engineer has to keep in mind is how to measure the geometrical accuracy of the tool. His task in this regard is made un-complicated by the existence of internationally agreed standards that prescribe the required specifications. His access to these standards and the fact that they exist, enable him, through comparison, to test the corresponding accuracy of his own tools. In other words, there is an independent reference point in relation to which his tools might be assessed, found wanting, or adequate in precision.

When we deal with cultures, however, there is no extra-culture standard which would enable us to opt for a cultural system on the basis of rational choice, or to judge another as inferior. Yet it is not so long ago that educated men were arguing in favour of a Second Genesis, the creation of all men in the image of Western man. The Europeanization of the world-picture (Barraclough's complaint) was repeated (and still is) in other areas of human experience.

Thus, a history of art turned out to be a history of European art,[6] and a history of ethics, a history of Western ethics.[7] Now as far as my knowledge goes, there has been no society that did not have a congruent system of ethics. As Professor Levi-Strauss wrote in *Race and History,* "men whose culture differs from our own are neither more nor less moral than ourselves: each society has its own moral standards by which it divides its members into good and evil".

Recently, Dr. Joseph Needham, sinologist, suggested quite bluntly that Western moral philosophers might profit from a serious study of how the Chinese went about inventing an ethical system that, unlike the Western one, was not based or dependent on a religious standard.

The assumption of Euro-American primacy in humanness has not been without effect on studies of non-Western societies and civilizations. I have used the phrase, "an imperialism of categories", to condense the meaning of the illegitimate examination and evaluation of non-Western societies through criteria fashioned in the Western context. For an analogous situation in technology discussions, I have spoken of the "tyrannization of historical possibility", that is, the refusal on the part of a great majority of learned men in the West, and their pale imitations in the Southern hemisphere, to accept that there are possibilities, theoretically defensible, of technological futures other than the ones they were indoctrinated with. Though, practically speaking, Guatemala, Greece, Chile, Vietnam and so on constitute proof that any choice of new societal directions may be subverted by big-power politics.

Thus it is through a critical assessment of the presentation of various intellectual traditions that I attempt to bring forth a ground-plan that is more realistically oriented to the exploding perspective of our times. Here, in a sense, Barraclough's task is easier than mine: the relativization of the places of Churchill or Hilter is hastened through the weaving of a larger net of human foci: Mao and Nehru and Nkrumah. All that is possible for me here is a re-instatement of Herder's conviction of the relativity of cultures.[8] For technology, my task is easier: there is enough evidence to establish a plurality of technical histories for the past and the possibility of different, alternative technological systems for the future.

Two final observations, one, on the context of this book, the other, on its style.

The reader may judge it paradoxical to learn that this volume was completed within the portals of a Western university, the Technological University of Eindhoven, in the Netherlands. The incongruity is more apparent than real, for nothing more is being attempted in this work (if argument is needed) than pushing the demands of Western scholarship to their rigorous limits. I have been brought up to believe that all Communist scholarly works are ideological, not objective, therefore untrustworthy. I have left these works aside for they are easy prey. But the reader will surely be surprised to find that a large majority of Western works have been equally ideological, and as this book will show, equally untrustworthy. The nature of objectivity is greatly transformed when we assume a world-wide perspective on human matters: in its new form, it makes large, nearly impossible demands. All the same, that is no reason why these demands should be refused.

I am not so sure, in the final analysis, whether this book would have been permitted in any university outside Dutch borders. The Dutch are like the Hindus in their great tolerance. Not only did they finance my bread-and-butter needs while I was in their country, they even awarded me a doctorate for my trouble, and the Department of International Affairs at the Ministry for Education and Science Policy contributed a substantial grant towards the printing costs of the University edition, from which this work is condensed.

Where the Dutch saw difficulty was the style in which this book was written. You may be as tolerant as a Hindu, but that still does not prepare you for a Hindu actually operating in your midst. Certainly, no Western scholar would have written a book in this manner. For myself, I confess to a culturally acquired disinclination to present a linear argument that runs all the way from the inception of a book to its terminus: I prefer to pick up a theme, and while discussing it, pick up another, and a third, a fourth, and so on. The final chapter is the jigsaw puzzle complete. So much for the wisdom that the Indian mind tends to work in ever widening spirals of thought. The unaccustomed reader is forced to pay greater attention as the book unfolds. Perhaps, here is his or her chance to make a small contribution to cross-cultural understanding. My compatriots, I think, will be infinitely pleased.

CLAUDE ALVARES

The first of January, 1978.

A Note on Terminology

The term *model* is used for a skeletal description of man, abstracted from the human experience of all cultures, emphasizing here his natural predisposition for creating culture and technology.

The term *paradigm* is not to be confused with the paradigms of Thomas Kuhn in *The Structure of Scientific Revolutions*. A paradigm here (to lapse into Plato's terms) is the *model* reflected in the concrete historical experience of any particular society. There is thus a Chinese paradigm or the Chinese edition of the *homo faber model* in culture and technology. There should be as many paradigms then as there are cultures. But there is only one *model:* the paradigms participate in it.

With the period of Western dominance, the Western paradigm became *identified* with the model and it consequently became possible for Western culture and technology to assume the role of a universalist standard in relation to which other paradigms could allegedly be evaluated. This book will attempt to separate once more the model from the paradigm and to propose the reconstruction of a situation akin to the one that existed before Western dominance: the independence of many different paradigms.

About *technology,* though I use the word very often, I should warn the reader that I have in mind a technological *system*. We live within one or another technological *system*. I do not accept the distinction between primitive, advanced, or intermediate technological systems.

I refuse also the terms "third world", "developing", or "poor" to describe the nations of Africa, Asia, and Latin America. Instead I have used a blanket term, Southern nations. The older terms define these Southern countries in relation to the industrial ones. This, I hold, is no longer necessary nor even desirable. Alternatively, I do not use the terms "developed", "traditional", or "advanced" in the text.

Elites: When I use the word elites, however, I use it to define power minorities in *all* societies, including the so-called democratic nations of the West. In fact, it is possible to argue that the elites of the industrial nations are more powerful than those of the Southern nations.

Introduction

THE POLITICS OF ANTHROPOLOGY

"Through the years," wrote the Argentinian writer, Jorge Luis Borges, "a man peoples a space with images of provinces, kingdoms, mountains, bays, shops, islands, fishes, rooms, tools, stars, horses and people. Shortly before his death, he discovers that the patient labyrinth of lines traces the image of his own face."

Borges, writing as a Latin American, would not have refused to extend his description of the life of a man to the life of a culture. For, through the centuries,. even a culture imperceptibly peoples its geography with the stuff of its own living creations. Natural objects, living beings, and all artefacts, come to be manipulated by a culture's invisible hands and to speak its silent language as they carry the imprint of its forms of expression, its thinking processes and philosophy, its language and technics. With the passage of time, a culture discovers in itself a distinct face, regulated and controlled by a distinct mind.

The American anthropologist, Clifford Geertz, in his recent volume of scintillating essays, *The Interpretation, of Cultures,* would therefore proceed to define culture as an internally consistent system of meanings. "Believing with Max Weber," he wrote, "that man is an animal suspended in webs of significance he himself has spun, I take culture to be those webs, and the analysis of it to be therefore not an experimental science in search of law but an interpretative one in search of meaning."[1]

Another anthropologist, Ruth Benedict, had early in this science's modern phase, already condensed the implications of such a view as Geertz's, when she had concluded that all cultures should be considered, in principle, with equal value.[2] This had been Herder's conviction already nearly two long centuries ago, when he had maintained that every society known to his time – whether Indian, Chinese, Egyptian, or Greek – had grown and developed in a distinctive manner and in response to the combination of environmental conditions presented by its particular time and place.[3]

Therefore, he noted, a society should be considered like an organism, in and for itself, "without foisting any set pattern on it". Human history may be better described not as a movement of different peoples towards some convergent mythical future (although at different speeds and in distinct groups), but as the experience of many discontinuous cultures, each in itself equally important as exhibiting the variability of the products of human inventiveness, each crystallizing a system of meanings irreducible to the others.

Thus Geertz reported that in Java, cultural opinion decreed: "To be human is to be Javanese." Other fields, noted the Javanese, other grasshoppers.

But the anthropologist is also a grasshopper, and thereby hangs a tale. The anthropologist rarely questioned the premises of his own culture: the very encounter with an alien community living quite creatively within the bounds of its own meaning system, should at least have cautioned him to reconsider his own "field" as merely one among many, all more or less fertile to the imagination of man.

I would be the first to acknowledge that the anthropologist was indeed a courageous and heroic person: more often than not, he had to struggle to live in the community he studied. And those laborious catalogues he compiled of genealogies, objects, myths, and vocabularies provide ample testimony to the sincerity of his purpose. But at the beginning and end of it all, he allowed the fact of his grasshopperhood to slip beyond the focus of his immediate task. If he documented, faithfully, the ways of life of the alien community, he did it within a framework of mind that located the community at a level lower than the one on which he himself, as a member of Western culture, stood and lived.

Such a situation favoured the proliferation of comfortable myths, the most persistent one of which taught that non-literate peoples in Southern countries possessed something like a "primitive mentality", that was not merely different from, but inferior to its Western counterpart. This "primitive mentality", the myth noted, was highly concrete, while the Western mind was more "abstract". The former was also supposed to connect its ideas by rote association, while the latter used general relations. Further, the nature of the Western mind disposed it to logic and consistency, while the primitive mind floated easily in contradictions: it was also generally more emotional and childish, and so forth.

It is a relief to discover, that though such theorizing has not entirely disappeared, it has at least been empirically falsified.[4]

Gerrit Huizer and Bruce Mannheim are editing a volume in which a group of anthropologists born in Africa have set out to dissect the "imperialism of categories" that determined the researches in that continent of a number of leading, non-African, primarily Western anthropologists.[5] Our point, however, is that anthropology is merely one example. The African anthropologists' analysis of the ethnocentrism supporting the main body of anthropological literature on African peoples can be repeated in areas as diverse as political science, economics, art, law, sociology, and psychology, more crucially, with the understanding of technology and culture in the world at large.

This submission of the vast panorama of human life to the rule of a monolithic paradigm is not a product of our own age. In fact, it first originated at the beginnings of the nineteenth century. Before that, something analogous had taken place *within* individual societies themselves. It is not necessary to prove, for example, that dominant classes and races have often held quite distorted views of those subject to or dependent on them.

Up to 1861, for example, the year the British colonial government introduced a more uniform penal code in India, a Brahmin murderer (under the existing Hindu law) could not be put to death. Yet a Shudra, a member of the lowest caste, who happened to sleep with a woman of a higher caste, usually on the latter's seducement, would automatically suffer execution. In China, as Etienne Balazs observed, traditional Chinese history was written "by scholar-officials for scholar-officials". A man who rebelled against the established order was termed *fei,* which is also a negative particle in classical Chinese grammar.[6]

In Britain, Karl Marx substantiated his critique of traditional political economy by pointing out that the classical economists, in accepting private property as *given,* had subordinated their science to the social interests of the ruling, landowning class against the interests of the classes subservient to it, a proposition firmly supported by the careful scholarship of Brian Inglis, in his extraordinary study on *Poverty and the Industrial Revolution.*[7]

These issues constitute the core of the larger discipline of the sociology of knowledge, into which we shall not enter here.[8] Our aim is a description of the situation that has come to obtain not between *classes,* but between *cultures,* more appropriately between dominant and dependent cultures. I am not being entirely original in suggesting that the political dominance exercised by the industrialized nations has made possible a parallel dominance of one cultural conception over and against all others unable to integrate themselves within its framework. Let me be concrete and turn to history for details.

In 1498, Vasco da Gama inaugurated the sea route to India. From that year, and for the three centuries that followed, Asia proved to have a larger and more powerful impact on Europe than is normally realized. Donald Lach has appropriately titled his first volume in his *Asia in the Making of Europe* series, as *The Century of Wonder*.[9] Later, we shall describe the material and cultural influence of Asia on Europe between 1500 and 1800. The current infatuation of the Southern nations for the West is of briefer duration.

The reversal of the European image of Asia seems to have occurred, however, in a gradual period between 1780 and 1830, by which time the foundations of the industrial revolution in England had already been laid. Voltaire noticed a bit of it. Having once considered India as "famous for its laws and sciences", he felt it necessary to denounce the increasing preoccupations of Europeans in India with the amassing of "immense fortunes", and this led him to remark that "if the Indians had remained unknown to the Tartars and to us, they would have been the happiest people in the world".

By 1830, the British had acquired, in what was to become a completely European century, a flattering notion of the nature of their own civilization, and a thoroughgoing contempt for every other. In India itself, this new attitude found expression in the famous Minute of Lord Macaulay on the 2nd of February, 1835:

> I have never found one amongst them [the orientalists] who could deny that a single shelf of a good European library was worth the whole native literature of India and Arabia. . . . It is, I believe, no exaggeration to say that all the historical information which has been collected from all the books written in the Sanskrit language is less valuable than what may be found in the most paltry abridgment used at preparatory schools in England. In every branch of physical or moral philosophy the relative position of the two nations is nearly the same.

Macaulay went on to note that the Board of Public Instruction would be wasting public funds should it print books of Indian learning "which are of less value than the paper on which they are printed was while it was blank", and that the artificial encouragement to "absurd history, absurd metaphysics, absurd physics, absurd theology" would end in the raising of a "breed of scholars, who live on the public while they are receiving their education, and whose education is so utterly useless to them that, when they have received it, they must either starve or live on the public all the rest of their lives". Little did Macaulay realize that it would be precisely the English system he introduced that would produce the "breed of scholars" so characteristic of India and the other Southern nations today: the educated unemployed.

That was, as I said already, in 1835. In 1959, the Dutch historian, Peter Geyl, had no different view of these matters.[10] In a lecture delivered at that time, Geyl considered it a stupid proposal "that we should prescribe extra-European history as a subsidiary subject to our best students", and he went on to express the fear that the extension of the secondary school curriculum to include Asian subjects could "only lead to a disintegration of intellectual life and to hotchpotch in the pupils' heads". In chapter one, we shall observe Jacob Bronowski moving along a similar rut.

The next influential person on our list, Karl Marx, had his own theories about the role of British industrial civilization in India. "England," he wrote, "has to fulfil a double mission in India: one destructive, the other regenerating – the annihilation of old Asiatic society, and the laying of the material foundations of Western society in Asia." He went on to emphasize how the British were breaking up the village community, uprooting handicraft industry, and establishing private property in land, which he termed "the great desideratum of Indian society". Industrial life would wreck the caste system: "Modern industry, resulting from the railway system, will dissolve the hereditary divisions of labour, upon which rest the Indian castes, those decisive impediments to Indian progress and Indian power."

Here, again, the remarkable fact is that a hundred years later, Peter Drucker (the godfather of the global corporation) would still be theorizing along similar lines. In one of his not so well known books, *The Landmarks of Tomorrow,* he urged his readers to face "the new reality of the collapse of the East, that is, of non-Western culture and civilization, to the point where no viable society anywhere can be built except on Western foundations". He based this pontification on the perception that:

> Every single one of the new countries in the world today – including those that have not yet shaken off colonial status – sees its goal in its

transformation into a Western state, economy and society, and sees the means to achieve this goal in the theories, institutions, sciences, technologies and tools the West has developed.[11]

Both Marx and Drucker, in formulating their predictions, were running away with the evidence. Predictions (which have a habit of turning into prescriptions) to be even mildly credible, must be preceded by at least a description of what *is*. And certainly no accurate description of non-Western human experience could ever have been possible with minds convinced, for example, that Western philosophy was the nearest approach to metaphysical truth ever attained by mankind, that the Christian religion contained truth incumbent upon all men everywhere to believe.

As Dr. Needham put it, even European painting and sculpture had become "absolute" painting and sculpture, that "which artists of all cultures must have been trying unsuccessfully to attain". European music was music, all other music, anthropology.[12] The study of white men, even, was a separate science called sociology: anthropology was for the rest. Methodologies of knowledge acquisition that ignored the scientific frame of reference were permanently unrealiable or ideological, to be studied properly as matters of "historical" interest, as the curious, quaint contents of a collection in the museum of mental history.

If Macaulay was one of the first to set out to prescribe how best, in his case Indians, might save their withered souls, he was also the virtual founder of a movement that would carry on his tradition to the present day. And the history of prescriptions concerning how non-Western societies might best achieve Western standards of life is studded with a long list of illustrious names, including Bertrand Russell, John Dewey, Talcott Parsons, and Ruth Benedict.[13] The last two scholars sat on a task-force charged with the problem of incorporating Japan within the American economic pattern, and Benedict's volume, *The Chrysanthemum and the Sword,* was commissioned by the Office of War Information.[14] Lately, we have had Project Camelot,[15] the revelations by Noam Chomsky concerning Vietnam,[16] and, of course, the entire disgusting family planning movement.[17]

THE POLITICS OF POLITICAL SCIENCE

The end of the Second World War was followed by the rise of a fresh generation of states, and by the time the world had entered the fifties, a growing concern with the phenomena of "backwardness" and "underdevelopment" had come to the fore. The leaders constituting the governments of the new states were faced then with the issues and problems that Latin American nations had faced earlier and African states would in the near future: planning for "development" and the choice of a suitable strategy to effect it.

These men of government, with rare exceptions, had been brought up in the Western tradition. What is not so well known is that they were not very experienced in the art of government. In such a situation, it was natural that the West, as inspirer, should soon turn into the West as guide. American experts sat on committees to formulate the First Indian Five Year Plan. And in 1960, Walt Rostow, in one of the most influential books of the decade to follow, set out to argue the credibility of their prescriptions.

Actually, as is now well known, Rostow's book, *The Stages of Economic Growth,* was not concerned at all with the "backwardness" of the new states, but formed part of a tactic designed to aid Dulles against Khrushchev in competing for the allegiance of these nations, still uncommitted to either of the two power blocks. The sub-title of Rostow's book, *A Non-Communist Manifesto,* underlined its nature of being an overtly political instrument of cold was ideology.[18]

Rostow argued that the key to successful development lay not with the Soviet Union, but with the West, that it was therefore in the interests of the non-aligned nations to jump on the Western bandwagon. W.F. Wertheim was one of the few Western thinkers to notice the nature of Rostow's description of Russian industrialization: that the American economist's famous thesis of a "take-off to sustained growth" provided an "argument for explaining away the specific significance of the Russian Revolution", by claiming that the "take-off" in the Russian economy had occurred twenty-five years before the revolution of 1917.[19]

It would take another fifteen years before scholars would isolate the fundamental deficiencies of Rostow's model;[20] by the time the critique had been accomplished, the economic and industrial elements of the Western paradigm, in so far as they might have had significance for the new nations, had lost their great appeal. But not before the serious flirtation on the part of the industrializing nations with the model had resulted in a powerful current of dismay, disillusion, and disappointment.[21]

Yet, Rostow was not an isolated example. In the area of political science, Western scholars had been equally busy constructing similar, only more bizarre, models and disseminating slanted advice. These scholars were not concerned with the problems of the new nations in their attempts to industrialize (that being Rostow's business), but to modernize.

The model presented by these scholars on a platter, so to speak, was again a distinctly Western one: formal democracy in combination with a rationalized bureaucracy; the new states should dispose themselves to attain this, since it represented the "summit" of political development. New states not yet incorporated within the model were to be termed "traditional", or better still "transitional", that is, still undergoing the throes of modernization. W.F. Wertheim again noticed the political implications and pointed out that the chief exponent of the school, Daniel Lerner, was guilty of extreme ethnocentrism in identifying the traits of modernity with those characteristic of American society.[22]

The modernization debate has its roots in the classical sociological tradition, a tradition which occupied itself with the construction of a conceptual model revolving round a pair of polar concepts, and then attempting to grasp through this the structural changes taking place as a "traditional" society moved into modernity. The polar concepts included pairs such as status and contract, *gemeinschaft* and *gesellschaft,* traditional and bureaucratic authority, sacred and secular associations, informal and formal groups and so on.

Talcott Parsons elaborated further on these to produce *his* own list of five pairs of alternative value-orientations or polar choices, which he claimed exclusively indicated the contexts of any social action whatsoever. Parsons had departed from Max Weber to the extent that his model was a wholly abstract one, not constructed through the study of any existing society. His five polar choices included:

affectivity versus affective neutrality
particularism versus universalism
ascription versus achievement
diffuseness versus specificity
self-orientation versus collectivity-orientation

From Parsons it was but a small step to contend, as Bert Hoselitz would do, that the developed Northern countries all exhibited the polar choices of universalism, achievement and so forth, while the

Southern nations were characterized by a total allegiance to the alternates: ascription, particularism, diffuseness, and self-orientation. A smaller step, again, for the description to turn into prescription: Southern nations need only eliminate their polar choices and opt for those characterizing Western societies. "Development" had become as simple and as exciting as a mathematical game: Kindleberger labelled this model the gap approach: you subtract the ideal or polar features typical of Southern nations from those of the Northern nations and the remainder is your development programme. Neat.

Our point, however, is not this obvious inadequacy, gross as it was. It is not difficult to prove that the godfathers of the gospel of modernization, including Lerner, S.N. Lipset, and Karl Deutsch, were influenced in their studies and policy recommendations concerning the Southern nations by categories and historical possibilities fashioned in their own context. And during the course of this book, we shall continually discover that an enquiry is immediately false if and when it sets out to ask whether a particular society (say, Guatemala) approximates the standards of the society (say, America) to which the comparativist belongs. In fact, it will always be misleading to attempt to discover whether a non-Western society has or can have or need have such characterological, structural, or philosophic features as an achievement ethic, modern bureaucracy, individualism, or an attitude of mastery towards nature as available supposedly in the industrial nations. And it is a relief to know that more and more social scientists do not accept the idea that the contemporary world (including the so-called new nations) is susceptible to meaningful social or political interpretation "by analytical schemes that dwell principally on twentieth-century economic phenomena".

There is a more serious criticism of the modernizers, available to us in the writings of the Indian political scientist, Rajni Kothari, namely, that the mode of development presented under the generalized package of the "modernization" process, undermined, in one continent after another, national independence in real terms, in the name of economic development.

Kothari inaugurated his fundamental criticism of the large body of literature produced by Western political scientists by pointing out that its overall perspective was always *apolitical,* for in its search for general principles, it ignored "the crucial problems of political reality, namely, the prevailing pattern of dominance and control in the world at a given time". The world was not simply divided between "developed" and "developing" societies, "modern" and "traditional" nations, but more importantly, between "dominant" and "dependent" powers.[23]

The modernization call to the Southern nations and the list of its prescriptions ignored this fundamental issue by directing attention to other tasks, including population growth, economic growth, family planning, literacy, development of media, and transfer of technology; the presumption being, of course, that once the basic social and economic tasks were performed, the necessary "political development" would follow: or, before states and nations are built, societies ought first to be "modernized". There was but one course of modernization open: a high rate of growth in GNP, increasing urbanization, a "rational" bureaucratic establishment, a manipulative technology, and "social mobilization" in which major clusters of social, economic, and psychological commitments are eroded or broken down and people made available for new patterns of socialization and behaviour.

The consequences of such an empty, context-free model of modernization had indeed been disastrous: it had produced an economic, bureaucratic, and technocratic elite intimately tied to the metropolitan areas of the world, treating the vast rural hinterlands in its own countries as colonies that provided cheap food and raw materials and surplus labour (and markets for inferior industrial products and obsolete industrial machines); an elite that had achieved high economic status at the expense of large numbers of people huddled in the "countryside", and in the process lost both its independence and its social conscience.

Here we are at the very heart of the "dependencia school" of thought, invented principally by the Latin Americans: the development of underdevelopment, as Andre Gunder Frank, its leading exponent, once put it.[24]

Thus, politically, Western-oriented political science reflected and reinforced, one might say, a division of the world in which large majorities of the societies politely labelled "nations" or "states", had in their entirety become the "countryside" and a small minority had in their entirety become the "cities", the metropoles. Further, the choices this science offered could now be seen to be false ones: they involved merely the possibility of political *forms* (democracy, dictatorship and so forth) or of political *games* (bargaining, coalition-making, power struggle etc.,) and not the required fundamental possibility of choosing preferred socio-economic goals and technological alternatives.

As Kothari began, in conclusion, to count the number of states (besides, of course, the industrial nations) which could be considered genuine polities where political autonomy and sovereignty were the distinctive features, and which had in the past even asserted their political autonomy and control of their economic futures and national security at the cost of reduced growth rates, he found precisely seven (one of which has already disappeared from the list): Yugoslavia, China, India, Rumania, Chile, Tanzania, and Cuba. And it is exactly with these nations that the prescriptions of the Western world had ceased to be worthily valued.

In the ultimate analysis, there is absolutely no reason for restricting the models of modernity and the processes and sequences of modernization to the experience of the Western nations. If, however, we continue to do so, we are easily open to the charge that we are subduing vast and varied societies to the totalitarianism of a single historical pattern. History might pattern itself on the past, but there is no reason it should pattern itself on the Western past; merely because, for example, Western nations realized conditions such as urbanization and literacy before political democracy is no proof that these are always prerequisites of it. Even for the purposes of a more wholesome science, it would be best to set no limits to the social and sociological imaginations.

Moreover, there has never been any conclusive evidence for the idea that Southern nations carry institutions and norms from their past that inhibit any increase, if necessary, in the efficiency of their productive systems. The opinion of the Dutch management theorist, P. Hesseling, is unambiguous in this matter:

> Any culture which has demonstrated survival value for a society over centuries is equally valid as any other culture which has proven its survival. Strategy translated into organization structure for the pursuit of complex work goals (products, service, or knowledge) is only a subset of cultural relationships. There is no evidence that a culture successful for a society would be incompatible with an effective work organization applying advanced technology that is consistent with the local or national culture.[25]

A great deal of evidence of such a point of view is to be found between the covers of books such as *The Modernity of Tradition,* by the Rudolphs, who set out to show how in India traditional structures and norms have been adapted or transformed to serve the needs of a society facing a new range of tasks.[26] The Rudolphs' study of the caste system, how its structural, cultural, and functional transformation has *aided* India's peasant society to make a success of democracy by enabling notables and parties to mobilize a popular vote or to achieve fundamental economic goals, goes directly against the common opinion that caste is an impediment to any serious improvement of the Indian situation. J.C. Heesterman, a Dutch indologist who has studied the inner conflict of value in Indian tradition, writes that "India has been remarkably successful in setting up the institutional framework for dealing with the traditional conflict in its modern reincarnation".[27]

I would be happy the day an Indian or African psychologist took up these modernization theories we have criticized for a deep analysis, for these theories undoubtedly tell us more about the nature of Western social scientists than they do about the actual situations in the Southern states. An analogous situation, for example, is American writing about China during the period of the cold war, which revealed more about American collective fears than it did about China itself. The Rudolphs provide a clue, which approximates issues normally dealt with in the sociology of knowledge:

> Africans, including American Negroes, long appeared to Americans as black, lazy, cannibalistic, chaotically sexual, childish, and incapable of social organization and government. We liked them that way because it strengthened the mirror image we had of ourselves as white, industrious, self-controlled, organized, orderly, and mature. India seen as a mirror image of the West appears otherworldly, fatalistic, unegalitarian, and corporate. It is as though we would be less ourselves, less this-worldly, masterful, egalitarian, and individualistic if they were less what they are. Occasionally one comes away from a colleague's work with the impression that he is reassuring himself and his readers of the uniqueness of the Western achievement, a uniqueness that would be endangered by recognition of the cultural, functional, and structural analogues to be found in non-Western, traditional societies.[28]

THE POLITICS OF PSYCHOLOGY

Within the "prescriptive" tradition represented by Lerner and Rostow, the most absurd ideas have come from David McClelland, who sees the problem of development as psychological, rather, as psychiatric.[29] For McClelland, the alpha and omega of economic development and cultural change can be totally identified as a high degree of individual motivation or *n*(eed) achievement, as the phrase has it:

> In its most general terms, the hypothesis states that a society with a generally high level of *n* achievement will produce more energetic entrepreneurs who, in turn, will produce more rapid economic development. . . .

Thus, the subtitles of the different sections of the final chapter of McClelland's *The Achieving Society,* which is devoted to "Accelerating Economic Growth", speak for themselves: what should be brought about in the Southern nations is the following list of necessities – Increasing Other-Directedness and Market Morality, Increasing *n* Achievement, Decreasing Father Dominance, Protestant Conversion, Reorganizing Fantasy Life, Utilizing Existing *n* Achievement Resources More Effectively.

> So we end on a practical note: a plan for accelerating economic growth through mobilizing effectively the high *n* Achievement resources of a developed country to select and work directly with the scarcer *n* Achievement resources in underdeveloped countries particularly in small and medium scale businesses located in provincial areas. . . .

Thus, the social and economic conditon of poor societies may be changed simply by having more of their members taught to get a hold of themselves and raise their need for achievement, or, as Andre Gunder Frank set out to parody it, "by having teachers and parents tell children more hero stories so that when the latter grow up they might be heroic developers themselves".

The most revealing part of the tale is McClelland's explaining away of everything worthy of China's attempts to emancipate itself, similar to Rostow's doctoring of Russian history: the Chinese have simply had more *n* Achievement and *n* Power than, say, the Indians. The role of all else is denied or devalued: the radical re-structuring of Chinese society, the influence of Mao Tse-tung, the revolution itself. Is this social science? S.N. Eisenstadt, in reviewing *The Achieving Society,* compared it to the work of Max Weber, to indicate to his readership "the measure of the importance of the problems raised by McClelland's endeavour". This is the first time I have come across a comparison that is really odious.

THE POOR AS SCAPEGOATS

It is the sequel to this "history of prescriptions" that is perhaps so expressly uninspiring. The empirical failure of all these prescriptions has been already obvious to anyone who has kept himself informed about the situation in the Southern nations: that for the majority of their populations, life has become materially harder and poorer. At first, attempts were made to find some consolation in the end result of all this prodigious activity: computing favourable aggregates and GNPs. Tirelessly, endlessly, comparative figures on national and per capita incomes were churned out to demonstrate that there had at least been some success stories: Venezuela, Lebanon, Pakistan, Liberia, South Korea, Thailand, and former South Vietnam – all of which had produced "miracle" growth rates. It was considered unnecessary for the moment to dwell too long on the discovery that these miracles had bypassed the basic needs and welfare of large sections of the population of these nations and further compromised their political independence in the polity of nations.

When the larger truth, however, was eventually accepted, it became all the more necessary to locate a scapegoat that had "frustrated" every well-meaning attempt at development. Excluded a priori were those experts that had prepared or advised the general strategies for the eradication of poverty: having laboured of their own free will in the shouldering of the white man's assumed burden, it would have been impolite to accuse them of error. On the other hand, it was equally embarrassing to accuse the intellectually bankrupt governments of most of the Southern nations for the continued, troubled state of affairs.

That left open that final, easily available, common target for the abuses of despair: the low-income groups, including the landless labourer, the small farmer, the unemployed craftsman. And since these could be calculated upon not to react or retun the attack, experts and governments set about the task with a will.

In the literature, it was Gunnar Myrdal who lent some sanction to the stereotype of the poor man as being "mostly passive, apathetic and inarticulate".[30] Others would follow him now (with a thesaurus in hand) to announce a list of unflattering epithets expressing "dismay": the poor man in the Southern nations, it seemed, positively refused to be helped out of his miserable condition. He was certainly a pathetic case to be pitied for his ignorance, stupidity, and unusual stubbornness in clinging to his traditional and illiterate ways.

Thus, George Coster, researching the underprivileged in Mexico, decried the "image of the limited good" that crippled their scanty aspirations.[31] Banfield, in southern Italy, though the principal obstacle "amoral familism", which in turn produced layers of "political incapacity" to demand and work out progress.[32] Charles Erasmus, again in Mexico, identified the way of life of the people there as being characterized by "inconspicuous consumption" and "keeping down with the Joneses", an attitude he found objectionable, even pathological.[33]

A few scholars in the European and American world of social sciences did make systematic attempts to run against the stream of inflated invectives. Gerrit Huizer, for example, painted a different picture of the underprivileged groups he worked with in Latin America and Italy.[34] W.F. Wertheim, in countless articles and books, expressed his conviction that the principal obstacles set in the path of the emancipation of the poor came, not from below, but from above, from ruling groups at the village, regional, national, and international levels, who only allowed change on their own terms.[35] Arjun Makhijani observed that in discussions about poverty and poor people, ruling groups often used the terms "illiteracy" and "ignorance" interchangeably, where they are manifestly different concepts, and then went on to suggest that the literate could do with some education to enable them to understand the point of view of the poor.[36]

But the popular mind continues to hang on to the hoary stereotype. That the tide of false perceptions has not been stemmed is evident in the influential opinion spread abroad that the poor, if they must remain poor, must at least be taught to stop breeding more of their kind. A variant of that legendary Marie Antoinette proposal has become the order of the day: if they have no bread, let them swallow pills.

THE SURVIVAL ENGINEER

Therefore, in a sense, this book.

There should be no ground for misunderstanding: this book carries no intent of defending the attitudes and perceptions of the poor, especially that large silent majority of low-income, permanently insecure groups in the Southern states. They need neither a defender nor a defence. If, on the other hand, this book lays any claim to originality, that might lie in its indication precisely of their *achievement,* the nature of which has less to do with the fact of their remaining poor than with their success in remaining alive.

Let me put it this way.

Consider a man born into a society as poor as India's. From the moment of his birth till that of his vastly hastened death, no welfare system will ease the tremendous odds in his path. If he needs a house, his lack of savings and credit will force him to build one from bits and pieces of cardboard or tin sheets. His life as a whole will follow a similar pattern, put together from a larger range of such bits and pieces. Unemployment will be always a near-fatality, not a social inconvenience as it is for his counterpart in the northern hemisphere.

More important, he is born into a society where power – political, economic, social – is exercised in the interests of protecting the privileges of just about five per cent of the population. He must sell his labour at a continuous disadvantage. If he owns land, he has little access to the means to improve it. And for the protectors and holders of privilege, he is little more than a burden, the coloured man's burden in a world where colour has made little difference to the nature of exploitation itself.

For those whose idea of what an engineer is, is restricted by the role of the engineer in highly industrialized societies, what comes now may seem difficult to accept. The traditional engineer is a person who uses trained skills and the insights provided by experience in the solution of technical or productive tasks. The engineer solves technical problems, and what is more the case today, exploits a vast arsenal of research facilities to create new technology.

This should not hide the fact, however, that the engineer, even at this high stage, is still an engineer of necessity: in a market that can only thrive on the principle of built-in obsolescence, the engineer must innovate or the productive system of which he is a part will not survive. Today, the threat of resource shortages has provided one more reason for him to innovate or invite disaster. Yet, if he loses, the worst that can happen to him is that he must join the ranks of the unemployed: the degree of his available comforts is decreased, but the welfare system sees that he still gets to eat.

The economically insecure man in the Southern nations is also, however, engaged in the task of survival, but this time, primary survival. Considering the range of odds against which he must struggle and his experience thus far in using all his wits about him to remain alive, he comes very close to being an engineer *par excellence*. The technology he uses is not invented for the maximization of profit; it is, instead, a *survival technology*, an expression used by Dutch philosopher Kwee Swan-Liat. Fully half the population of today's world are survival technicians; they do not exploit the Western technological system. They are craftsmen of necessity, and that necessity is in a very real sense rationally engaged.

For, a survival technology only makes sense if viewed in the context of a *survival algorithm:* a set of rules of farming, for instance, that ensure a tolerable minimum output.[37] Faced with an absence of capital resources, and a heavy dependence on climatic conditions, the farmer's principal focus is directed to the *minimizing of risk;* in economic terms, he prefers to give up some expectation of profit so as to reduce risk. A new, alien cropping system, because it has no history of experience to back it, contains a large component of risk; its failure means ruin, forced sale of land or debt.

A rejection of new ways in preference to the old is the protective reflex of the overwhelming majority of farmers who know very little except that life is a fragile possession, and that tried and true ways, however burdensome, have at least proved capable of sustaining. A mulish perseverance in old ways is not without reason when life is lived at the brink of existence where a small error may spell disaster.[38]

Thus, each family finds, inherits, and defends from experience, farming procedures that ensure survival as a landowning unit. This survival algorithm suits the family's risk aversion, preference between income and leisure, liking for various seldom-traded vegetables intertilled with the main crop, auspicious days and *dharma* (caste duty). In a complex affair that involves life and death, it would be inviting danger needlessly to accept a new practice that is "improved" only in the context of someone else's algorithm.

In the grand Chinese tradition that recommends a picture to a million words, here is an example: Margaret Mead once wrote of a team of United Nations' agricultural experts who tried some years ago, in vain, to persuade Turkish farmers to improve their crops by removing the stones from their fields. Eventually, some of the younger farmers agreed. To the experts' chagrin, their yields promptly *declined.* In the arid climate of Turkey, it seems, the stones had served the crucial function of helping to retain scanty moisture in the soil.[39]

CHAPTER ONE

A New Anthropological Model

> For three thousand years a dialogue has been going on between the two ends of the Old World. Greatly have they influenced each other, and very different are the cultures they have produced. We have now good reason to think that the problems of the world will never be solved so long as they are considered only from a European point of view. It is necessary to see Europe from the outside, to see European history, and European failure no less than European achievement, through the eyes of that larger part of humanity, the peoples of Asia (and indeed also of Africa).
>
> –Joseph Needham: 1969

The fact that nearly 200 years after the industrial revolution first gained momentum, half the world's population still ties itself down in its struggle, for primary necessities, to survival technological systems is embarrassing enough. What disturbs one, however, is the idea, in the back of most learned men's minds, that this underprivileged part of the world cannot hope to save itself except through the replication and extension of the Western paradigm of cultural and technological development. The basis for this belief is the specious assumption that there is but one form of technological development, the most advanced and the best, that which came to fruition in the history of the Western world. The implication of such an assumption is worse: the discounting of any capacity in the Southern nations to solve their own technical problems.

Therefore, this chapter, which is an essay on the philosophy of technology and culture.

"If you wish to converse with me," said Voltaire, "define your terms." Our terms, *homo faber*, are Latin ones, and even M. Voltaire would have conceded that a line of thought expressed in one language undergoes a loss of meaning when translated into another. But the translation is easy. *Faber* in Latin means a smith or worker in hard materials, and a *faber* worked in a *fabrica*. The verb, of course, is *facio*, to make, and so *homo faber* means, quite simply, man the maker, which for our purposes is both inelegant and vague.

We need a term or phrase which, while remaining simple, would agree to contain a large reservoir of interconnected meanings, including those identified with man as he makes, builds, constructs, invents, fabricates or creates his world. We are interested in artefacts, and these are not to be restricted to tools and machines. Languages, thought systems and symbols are also artefacts, for they sprang into existence only after man began to walk the earth. Man extinguished, artefacts fall, meanings disintegrate, only natural events remain. There is no English word for this wide range of meanings, so the living having forsaken us, we stick to dog Latin and *homo faber*.

The English historian, Thomas Carlyle, called man a "tool-using" animal. The American, Benjamin Franklin, more confident, enlarged that to "tool-making animal". Not much of an improvement, yet the late Jacob Bronowski, who should have known better, was willing to be impressed. In *The Ascent of Man,* reviewed favourably even in India, he wrote:

> We have to understand that the world can only be grasped by action, not by contemplation. The hand is more important than the eye. We are not one of those resigned, contemplative civilizations of the Far East or the Middle Ages, that believed that the world has only to be seen and thought about and who practised no science in the form that is characteristic for us. We are active; and indeed we know, as something more than a symbolic accident in the evolution of man, that it is the hand that drives the subsequent evolution of the brain. We find tools today made by man before he became man. Benjamin Franklin in 1778 called man a "tool-making animal" and that is right.[1]

It is a comfort to know that so clever a man could talk such nonsense. Surely the arrival of numerous volumes on the science and technology of China, India, Africa, and pre-Columbian America, and the equally impressive literature produced by ethologists should have enlarged the late scholar's education, unless, of course, in accepting a British home, he embraced its insularity as well. Otherwise it is difficult to understand how he could continue to hold that man could be distinguished from animals because of his tools, or, by the same criterion, Western man from the rest of the world.

In her splendid introduction to Louis Dumont's *Homo Hierarchus,* Mary Douglas observed that "it is defeating to restrict sociological inquiry to modern industrial societies and so to restrict the very idea of what forms man in society can take". She went on to add:

> If we seek objectivity, we must recognize our own fundamental assumptions for what they are: the creation of our place and time. If we mistake our current idea of the nature of man for the eternal laws of nature, we are blinkered by cultural restraints on perception.[2]

Thus, the contemporary obsession with technology has led to attempts to pass man off as a "tool-making animal". One great philosopher of our times, however, has stood resolutely against the stream, and it is to him that I now turn.

The immediate provocation for Lewis Mumford's double-decker, *The Myth of the Machine,* was in fact the nearly universal description of man as a tool-using-making animal. In the prologue to this work, he cautioned against overstressing the role of tools in early man's development just because of man's obvious need for tools. Otherwise, he wrote, we ignore a wide range of other activities in which many other species have for long been more knowledgeable than man. Insects, birds, and mammals, for example,

> have made far more radical innovations in the fabrication of containers, with their intricate nests and bowers, their geometric beehives, their urbanoid anthills, and termitaries, their beaver lodges, than man's ancestors had achieved in the making of tools until the emergence of *homo sapiens*. In short, if technical proficiency alone were sufficient to identify and foster intelligence, man was for long a dullard, compared with many other species.[3]

The point that Mumford is making here (and one which Johan Huizinga also made) is that the narrow description of *homo faber* does not immediately serve to distinguish man from animal, from *animal faber*. Kohler's apes have done a great deal, together with modern-day ethology, to blur the distinction. That animal capacity has long been underrated is now obvious to those who follow the tremendous research being put into animal nature (recently summed up in a volume by W.H. Thorpe).[4] Niko Tinbergen notes:

> It was said that 1) animals cannot learn; 2) animals cannot conceptualize; 3) cannot plan ahead; 4) cannot use, much less make tools; 5) it was said they have no language; 6) they cannot count; 7) they lack artistic sense; 8) they lack all ethical sense.[5]

Tinbergen himself feels all such distinctions have either been dropped or come to be seen by ethologists as differences of degree and not as aspects of a fundamental discontinuity. Mumford suggests, in conclusion, that should we really need proof of man's genuine superiority to his fellow creatures, we would do better by looking for a different kind of evidence that his poor stone tools alone:

> Or, rather, we should ask ourselves what activities preoccupied him during those countless years when with the same materials and the same muscular movements he later used so skillfully he might have fashioned better tools.[6]

Suggestions supported by anthropologist, Geertz, in his recent volume. Beginning with the observation that most of the available evidence from archaeology and palaeontology firmly places the Australopithecines within the line of the hominids, and that the brain of *homo sapiens* is about three times as large as that of the Australopithecines, he concludes that the greater part of human cortical expansion has *followed,* not preceded, the beginning of culture.

In other words, it makes more sense to believe that "culture was ingredient", and that too centrally ingredient, in the production of the human animal, rather than to think of it in terms of being added on, so to speak, to a finished or nearly-finished animal. And by culture, Geertz has in mind much more than the mere perfection for tools. It also includes the adoption of organized hunting and gathering practices, the beginnings of true family organization, the discovery of fire, and most critically, "though it is as yet extremely difficult to trace it out in any detail, the increasing reliance upon systems of significant symbols (language, art, myth, ritual) for orientation, communication, and self-control". All created for man a new environment to which he was then obliged to adapt.

The restrictive theory that tool-making once drove the evolution of the mind is now laid to rest:

> Because tool manufacture puts a premium on manual skill and foresight, its introduction must have acted to shift selection pressures so as to favour the rapid growth of the forebrain as, in all likelihood, did the advance in social organization, communication, and moral regulation which there is reason to believe also occurred during this period of overlap between cultural and biological change.[7]

HOMO LUDENS, HOMO SYMBOLICUS

Besides *homo faber*, a generalized philosophical anthropology offers us two further characterizations of man, *homo ludens* and *homo symbolicus,* and since *homo faber* is a layer of meanings that cuts across these other two, I shall pass to them briefly in turn.

It was the Dutch historian, Johan Huizinga, who first proposed the *ludens* concept, during his lectures over the world between 1933 and 1937. In the foreword to his volume of a similar title, which speaks for itself and the theme of his book, he wrote:

A happier age than ours once made bold to call our species by the name of Homo Sapiens. *In the course of time we have come to realize that we are not reasonable after all as the Eighteenth Century, with its worship of reason and its naive optimism taught us; hence modern fashion inclines to designate our species as* homo faber: *Man the Maker. But though* faber *may not be quite so dubious as* sapiens, *it is, as a name specific of the human being, even less appropriate, seeing that many animals too are makers. There is a third function, however, applicable to both human and animal life, and just as important as reasoning and making – namely, playing. It seems to me that next to* homo faber, *and perhaps on the same level as* homo sapiens, Homo Ludens, *Man the Player, deserves a place in our nomenclature.*[8]

Homo Ludens translated as Man the Player is diseuphonious, and like our own title, the original Latin was preserved in the English version of Huizinga's work. What robbed the book of its original intention, however, was the translation of the subtitle. Huizinga, in London, had subtitled his lecture, *The Play Element of Culture.* The English insisted (and finally the translator too) that it should be changed to read *The Play Element in Culture,* which, as Mumford was quick to point out, was tame and unrevolutionary, because imprecise – it did not at all express Huizinga's real claim: that culture is play, stark, and simple. After a wide-ranging study of various cultures, including China, he had concluded:

Civilization is, in its earliest phases, played. It does not come *from* play like a baby detaching itself from the womb: it arises *in* and *as* play and never leaves it.[9]

I have a large doubt whether Huizinga's description of man would still be useful to us in our understanding of human behaviour in industrial societies today. I shall argue, later, that industrial man was compelled to permit this important dimension of his life to suffer a temporary eclipse. A high-consumption society is, in my mind, incompatible with the play element of its culture. And the American dramatist, Walter Kerr, was not the first to document "the decline of pleasure" in the industrial civilization of which he was a part, but his analysis goes deeper than any other I have ever known.[10] Kerr underlined the wide influence of utilitarianism in American culture and the resulting inability of American man, at least, to play. In cultures still outside the reach of the industrial umbrella, Huizinga's *Homo Ludens* still thrives.

As important as play or tools is the symbol in the life of man, and this dimension of the human world was studied in comprehensive detail by Ernst Cassirer in his monumental work, *The Philosophy of Symbolic Forms*.

Cassirer noted that man no longer lives in a physical universe: he inhabits instead a symbolic world. Language, myth, art, and religion form parts of this non-physical world. In other words, these are the diverse threads out of which the symbolic net is woven and which sits in turn like a complex web on human experience. All progress in human thought and belief refines and reinforces this web.

Man, said Cassirer, is unable any longer to meet reality directly: between the real and the human mediates the symbolic. To the extent that the symbolic activity of man increases, the significance of the physical inversely decreases or recedes. The multitudinous world of languages, of images in art, of mythical symbols, and of religious rites, betrays the fact that man cannot know or encounter anything except through the medium of this pervasive edifice.[11]

The reason I have stated that *homo ludens* and *homo symbolicus* may serve as the foci of a generalized, transcultural, philosophical anthropology is because, as with *homo faber,* there is no culture which does not manifest their presence. On the other hand, *homo equalis* for example, may be considered a perfect sample of a particularized philosophical anthropology (in this case, Western), simply because in another philosophical anthropology, that of India, precisely the opposite characterization of man obtains, based on reverse assumptions, and which Louis Dumont has termed *homo hierarchus*. Anyone who sets out to apply the qualities of *homo equalis* to his understanding of Indian man is then automatically betraying his ethnocentrism.

A NEW MODEL

The older definition of *homo faber* merely in terms of tool-making ability is inadequate, since this identification ignores more than it uses. Further, the insights of Huizinga and Cassirer should also form essential elements in any realistic understanding of man. I shall now propose a new model, which I hope will take all these points into consideration.

This new *homo faber* model will also serve to distinguish man from animal, in terms of a figure-ground relation, in which man is set against animal, so his figure lies heightened against the foil provided by his animal past. I must not be misunderstood. I am wholly at one with Niko Tinbergen's conclusions above. Our "fundamental distinctness" from the animal world, I am concerned to note, because doubtful in theory, has led to inhuman practice: it has allowed us, with impunity, to make extinct entire species and to inflict brutal and senseless tortures on animals in our uncontrolled urge for doubtful scientific research. Every year, in the United States alone, at least 63 million animals undergo "experimentation". In England, the number is 5 million. These are merely two of the numerous countries with such programmes. The occasional twinge of conscience has dictated the "de-vocalization" of some animals, to prevent their screaming in pain. Cats have their paws cut (recommended) to prevent them from scratching in rage. And tests like the Draize formula, to try out new cosmetics, are carried out by dripping concentrated solutions of the product into the eyes, usually of rabbits, to measure the resulting injury.[12]

The new model proposed, elements of which come from the writings of the Dutch philosopher, Kwee Swan-Liat, and American anthropologist, Clifford Geertz, enables us to see, not just man, but animals too differently. For, if it helps us to distinguish *homo faber* from *animal faber,* this is because it proposes that any such differentiation is possible precisely because of a *similar* base in animal and man; and further, because it sees the cultural pluriformity of man again as a fundamental outgrowth from a structure that is very *similar* in both animal and human natures.

There are two parts to the model, one which measures the differing responses of animal and man in relation to culture, and the other, in relation to technology. We begin with the first.

Kwee writes:

Man is not alone in being able to construct an artificial environment. Animals too produce all kinds of artefacts. They produce, collect and work different types of materials, through which they are able to give an appropriate form and shape to their direct environment. Coral, spiders, bees, termites, birds, beavers: all are capable of impressive constructions. There are, however, important differences between the building activities of animals and those of man.[13]

The structures animals build, writes Kwee, are directed by an innate programme. Spiders, bees, and birds have no conscious goals other than their webs, hives, or nests. Yet, they are able to carry out, in proper order and coherence, faultlessly, purposefully, and effectively, all the various constitutive stages of their programme.[14]

On the other hand, man lacks this internal instinctive programme. But it is precisely due to this deficiency that he possesses the potential for learning from and through experience. On dissecting this experience, he is able to develop a continuous insight into the possibilities of his artefacts. Combining these possibilities within reflection, he is able to propose objects or goals far removed from the given.

Since the behaviour pattern of animals is internally programmed, they encounter difficulty in being able to choose or change or vary their activity. The pattern and the resulting structure are characteristic for the species, not for the individual within the species. The nests of swallows and weaver-birds may vary among themselves, but within the species themselves, there is little variation. The building programme or blueprint is already given: the implementation is instinctively regulated.

As we said earlier, because man does not possess a genetically determined building programme, he is able to invent a building programme himself. The plan, further, can be pictured, discussed, modified, and carried out (or kept in cold storage), either by the individual or through a social consensus. Stone-henge, the pyramids, palaces, temples, caves, the olympic stadia, cathedrals and supermarkets, as examples, are only sensibly understood within the framework of the goals and ideals that lie underneath. The goals may be described variously: they may be military, religious, political, or economic. Even playful.

This is by far the positive side, one side of the medal. For, when one reads Geertz, one discovers that man, like the animal, is controlled by an implicit interiorized programme. Geertz points out that man works out his life within certain limits, the limits set by his culture's web. As he puts it, the fact that culture was ingredient in the early development of man, means that while we have been obliged to abandon the regularity and precision of detailed genetic control, we have also simultaneously moved in the direction of a more generalized system of control provided by cultural sources, the accumulated fund of symbols. This symbolic control over our conduct is no less powerful than the earlier, now abandoned, instinctive programme:

> To supply the additional information necessary to be able to act, we were forced, in turn, to rely more and more heavily on cultural sources – the accumulated fund of significant symbols. Such symbols are thus not mere expressions, instrumentalities, or correlates of our biological, psychological, and social existence; they are prerequisites of it. Without men, no culture [Kwee]; but equally, and more significantly, without culture, no men.[15]

The same coin is being examined, but from different sides: there is no contradiction involved. There is no doubt that man is often free to invent and implement his own programmes. Kwee's ideas would find greater elaboration in the writings of the Spanish philosopher, Ortega y Gasset, who observed that from the point of view of bare living, the animal is perfect and needs no technology. In words paralleling those of Kwee, he wrote:

> In the vacuum arising from the transcendence of his animal life man devotes himself to a series of non-biological occupations which are not imposed by nature but invented by himself.... Is human life in its most human dimensions a work of fiction? Is man a sort of novelist of himself who conceives the fanciful figure of a personage with its unreal occupation and then, for the sake of converting it into reality, does all the things he does – and becomes an engineer?[16]

Yet, with Geertz, it is difficult to deny that the very structure of our thinking about goals, the framework of mind, is provided by culture. And because we have different cultures, we have in reality also different cultural paradigms, different programmes. If technology is an element of a total cultural system, if the hand is guided by the mind, one culture's understanding of technology will differ from another's. Therefore, in a very real sense, it becomes possible to speak in terms of alternative technological systems. There is a characteristic passage in Geertz that might have also been written by Kwee, with a turn of phrase:

Beavers build dams, birds build nests, bees locate food, baboons organize social groups, and mice mate on the basis of forms of learning that rest predominantly on the instructions encoded in their genes and evoked by appropriate patterns on external stimuli: physical keys inserted into organic locks. But men build dams or shelters, locate food, organize their social groups, or find sexual patterns under the guidance of instructions encoded in flow charts and blueprints, hunting lore, moral systems and aesthetic judgements: conceptual structures molding formless talents.[17]

Homo faber is also *homo fabricatus*. The ideas, values, acts, even emotions of men, like their nervous system, are cultural products, manufactured out of the capacities, tendencies, and dispositions with which they are born: men, like their cathedrals or their temples, are cultural artefacts.

The second element of Kwee's *homo faber* model concerns technology, and as with the first, Kwee discusses this issue by distinguishing animal technical capacity from human technological possibility. He writes, for example:

This important difference between the internal, genetically and instinctually programmed blueprint available to animals and the external, blueprint programmed by men through social communication and consensus connects with another fundamental difference, namely, that between the organic, bodily specialization of animals and the organizational and technical instrumentality available to man.[18]

In other words, both animals and men have reached a stage of bodily specialization, that is, the stage where their morphological development in evolutionary terms can be considered to be complete.[19] In the case of the animal, however, the state of specialization determines its activity: and in the case of man, the state of specialization itself provides a flexibility that allows in turn the fulfilment of purposes other than those internally programmed.

Thus spiders and caterpillars possess glands that produce material for their webs and cocoons. Colonies of bees and termites avail of a division of labour based on a bodily-determined specialization of functions. On the other hand, the hands of man have, very early, provided him with an escape-hatch through which he has been able to avoid the limitations of a set, morphological structure. These hands enable him to work materials, to build, and to construct according to a programme he himself constructs. Every culture has also a system of technology and, therefore, also a history of technology.

Yet, here too we encounter another perspective, another side to the coin. The technical possibilities of man are available within limits, the limits of the natural environment. A culture's ecology is a subsystem of a larger natural ecology that contains it. Material culture, the technological system, constitutes the bridge between the two.

The earth, in other words, comes before *homo faber:* the natural environment existed before he did. If he moulds it with the unique flexibility of his hands, it in turn moulds him as much. The course of civilization, as Will Durant put it, wends its way not necessarily westward, but in the direction of resources and fields. As man starts from the Tropics, the path of empire is mostly north and south; and today it may laugh at all formulae and turn backward to the east. But everywhere the culture of the soil precedes and conditions the culture of the mind.[20]

There is here, as before, no contradiction. If most species may only operate within the confines of their specific ecological niches, man's ability for culture enables him to create his own ecological niche or exploit different ones. Though in one sense, he finds himself restricted in having to adapt to his environment, in another it is his very ability to do so that lays the basis for a varied history of technology.

Few historians have been able to combine in their works, much less in theory, the peculiar combination of personal (Carlyle's Great Men thesis is a subset of this) and impersonal (socio-economic, ecological) elements that produce the vast drama of human life. If the great Chinese emperor, K'ang-hsi, for example, had succumbed to smallpox as a child, the course of Ch'ing history might have been very different: he survived, and did more than anyone else to consolidate the empire and dynasty of the Ch'ing.

On the other hand, if Hung Hsiu-ch'uan had not had a dream which proved to him that he was indeed the younger brother of Jesus Christ and thus inspired him to found the Heavenly Kingdom of Great Peace, then millions of people might not have died, and the Ch'ing achievement of K'ang-hsi might not have received the irreparable blow it did.

It is instructive to examine the rise of technology in places where environmental pressures prescribe harsh and strict limits to production. In *Craftsmen of Necessity,* Christopher Williams provides a panoramic view of a number of indigenous communities today that must cope with difficult temperatures and limited materials, and whose technology is in fact a response to environmental pressures.[21]

Williams notes that the people of his book have "concepts of form and function that have been developed, through experience, out of the biology of their land". What the observer will encounter is not complexity, but sophistication: the two should not be confused, as Robert Spier has pointed out. Complexity in the modes of production is more often a distinct response from a society where numbers have arrived at the threshold of importance:

> At some point along the line of increasing complexity of operations the law of diminishing returns sets in: the effort to possess and utilize all of the paraphernalia of modern life becomes too great for the returns received.[22]

The principles governing the lines of sophistication are not necessarily the same as those involved in complexity. They are results of two different existential situations. To categorize the former (because non-industrial) as belonging to a "lower" state of technical development and the latter as "advanced" is to betray a preference for technological development *per se,* a preference based itself on a belief, assumed of course, that technological change can be studied independent of the needs that have brought it about. Such a view, that practically parallels the orthogenetic view in evolutionary theory, is probably evidence that some scholars' minds tend to move in straight lines, not that technological change ever does.

Man, wrote Kwee, is morphologically disposed to technology and culture. He is, at it were, biologically outfitted for the use of tools and plans. Every culture brings forth architects and builders, craftsmen of necessity, poets, and entertainers. A culture without a system of technology is a contradiction in terms. There are no "poor" nations or societies: merely large masses of people whose ability to produce more than what is necessary for primary survival has been drastically reduced by external impediments. In the absence of such forces of suppression, technological change is natural, as new insights into the productive system result from the "research" of years or centuries of practical use and test.[23]

A hostile environment may be either political or natural, but its end result is to reduce the wide range of human ecological niches that culture provides normally to human capacity. In other words, human capacity is often transmuted into a mere ability to cope. Which means, as Arjun Makhijani has noted, that poor people using survival technologies make optimum use of their resources, because they are forced to do so: they often waste the least. Every bit and piece of their resources, ability, and wit is exploited to meet the demands of primary survival.

The other side of the coin, the limits to growth, has only been receiving some attention in recent years.[24] Before *Limits to Growth* appeared on the scene, most of us continued to be told in numerous volumes that industrial man had invented a new technological environment in which to move and have his being, which reduced his dependence on the unpredictable movements of nature.

The 1972 Report of the Club of Rome startled this complacent world. The document studied five factors: population, food supplies, industry, resources, and pollution, to conclude that mankind was headed for disaster unless it worked immediately in the direction of zero population and zero economic growth. In other words, the Club of Rome document pleaded for a return to ecological equilibrium on a global scale.

The relationship between a population, its productive system and the resources available to it is a delicate one, and if ecological equilibrium is upset for any reason, resource scarcity invites increasing difficulties in coping. At first, such a scarcity may be overcome by imports to cover the deficiency, through trade flows in general, including specialized production for export. At other times, a society may be forced to change its resource base entirely, look for new sources or develop more intensive methods of exploiting available ones. These will inevitably imply more involved, expensive, and complex production and processing systems. The substitution of coal for wood, for example, as forests dwindled, was not merely a matter of substituting a pick-axe for an axe: it meant all the problems that come with mining and the processing of coal for industrial and domestic use, besides the new threats to health.

As a rule, technical change that is necessitated gives rise to more problems than it solves. Mineral resources, for example, are less easily utilizable than land-based ones, like wood, and transport becomes increasingly a problem as local self-sufficiency breaks down. Transport adds only to the cost of goods, not to their value.

In our contemporary world, societies are pressing forward their nuclear energy programmes on the ground that it is the only possible alternative they have in a world faced with ever increasing depletion of fossil fuels. Those who value nuclear technology in this light will tend to highlight man's continuous ability to meet challenges, each more difficult and obnoxious than its predecessor. Others might be more disposed to see the contemporary rush to nuclear technology, even after realizing that many problems still remain unsolved, as the plight of societies with their backs to the ecological wall. A situation that more often than not carries the germs of war. The Japanese economy would collapse within months if cut off from resources for its productive system, most of which come from without. And the United States has ninety-two military contracts with an equal number of nations to protect its resource routes.

THE USES OF THE NEW MODEL

For the past fifteen decades, particularly during the last three, the peoples of Asia, Africa, and Latin America were told a different version of the story. Further, they were taught, directly and indirectly, to compare their technological systems in terms of the Western production system, and to define themselves and their cultures in relation to a very particularized philosophical anthropology. Every aspect of the life of their societies was then compared, judged, or assessed in terms of what obtained in the West.

Nowhere is this *attitude* (there is less *reason* to it than normally supposed) – that the West, for example, had produced a system of technology so "advanced" that it would replace all the others – more easily studied than in the literature, especially that which has arisen around the history of technology itself. And there is no better way to illustrate the general purport of this book than a brief analysis of the histories of technology that are behind us.

We might begin with the Dutch historian of technology, R.J. Forbes, who first published his history of technology and engineering, *Man the Maker,* in 1950. He used the term *homo faber* in the narrow sense I have criticized above, but he did make his position clear. Not unlike Mumford, he observed that early man assumed "the role of both craftsman and engineer in addition to that of artist, philosopher and teacher". However, he would use the expression *homo faber*, he continued, to denote "a sociological species distinct from *homo sapiens";* as it finally turned out, this "sociological" description preserved itself for a few pages, then gradually fused into a total description, in the same manner in which Ralf Dahrendorf's *homo sociologicus* would later cross over from model to real world.

Now, Forbes was one of the first historians of technology to conclude that technology was the work of mankind as a whole, and that "no part of the world can claim to be more innately gifted than any other part". Here, he was providing a useful corrective to the opinion of historian Arnold Toynbee, who had written earlier, on the basis of what evidence I do not know:

> However far it may or may not be possible to trace back our Western mechanical trend towards the origins of our Western history, there is no doubt that a mechanical penchant is as characteristic of the Western civilization as an aesthetic penchant was of the Hellenic, or a religious penchant was of the Indic and the Hindu.[25]

Four years later, Forbes produced the rich and prodigiously detailed, multi-volumed, *Studies in Ancient Technology,* which set out a remarkable-for-his-times description of the different technologies of Asia, Africa, pre-Columbian America, and Europe. But, it was with *The Conquest of Nature* that Forbes gave in to the attitude I have tended to criticise throughout this book: there he subsumed the numerous and varied technical acts of men in different cultures under a philosophical anthropology that was unmistakably Western: the domination of nature. Incredibly, the slim book even ended on the promise of our redemption from the consequences of faulted technology through the event of Easter!

Another influential work of the time was the German scholar, Friedrich Klemm's *A History of Western Technology,* which provided a picture of Western technological development in which non-Western technologies had no hand at all. Klemm's volume appeared in the same year as the first volume of Needham's work on Chinese technology. The English translation appeared, however, in 1959; in it, Needham's work is mentioned in the bibliography, for decorative purposes. This is obvious, because Klemm could only substantiate his interpretation of Western technological development by consciously playing down non-Western technics. In fact, the only quote on Chinese technology in the work is from the *Kwan-Yinn-Tzu*, the work of a Taoist mystic of the eighth century A.D., and it is paraded to prove why, in China, the religiously coloured, oriental rejection of the world could not have provided a stimulus for science and technology in that country.

My final illustration is, of course, the once-standard *A History of Technology*, edited by Charles Singer, E.J. Holmyard, and A.R. Hall in a series of five volumes. Though the first volume appeared in the same year, again as Needham's, and though the writers themselves acknowledged that up to the Middle Ages Chinese technology was the most sophisticated in its fund of technical expertise, little of Chinese technics was documented. Three of the Singer volumes dealt with pre-industrial technology, where China should logically have been given the major space: the Western technological development should have been added on perhaps as an appendix. Matters have not changed much since the volumes first appeared. Derry and Williams later condensed the five volumes into a single *Short History of Technology: From the Earliest Times to A.D. 1900.* They admitted knowledge of Needham's work, then went on to ignore it.

Singer's *History* has by now been laid to rest by Western historians themselves on one issue alone: that it studied technology in isolation from numerous other elements that had something to do with it. Such a treatment, wrote Melvin Kranzberg,

> derives inevitably from the definition of technology as "how things are commonly done or made" and "what things are done and made". But many other questions immediately come to mind: *Why* are things done and made as they are? What effects have these methods and things upon elements of human activity? How have other elements in society and culture affected how, what, and why things are done or made? The five volumes of *A History of Technology* codify the present state of scholarship, but they do not answer these further questions.[25]

In opposition to that, Melvin Kranzberg and others founded the Society for the History of Technology in the United States in 1958, which a year later began to publish a new international quarterly, *Technology and Culture*. The purpose of the Society was to study the history of technology in its relation to society and culture. Fair enough, admirable. Kranzberg further attempted to distinguish his new journal from others in the field, like the British Newcomen Society's *Transactions* and the French *Documents pour l'histoire des techniques,* by declaring that it would be the first truly international journal of its kind, "serving the needs of scholars in America and throughout the world".

A careful study of the journal issues over the past fifteen years, however, shows no great indication of Kranzberg's promised internationalism. The majority of the articles published are still devoted to Euro-American technological history and culture and the journal as a whole has not succeeded in avoiding the parochialism of Klemm.

There is a good reason for this continuing restricted approach to the study of technology. It is taken for granted that the technology that evolved in Western civilization is the only important one there *was,* in light of the present. We are, as I noted earlier, speaking here of an attitude which would make very little sense without a theory to back it. The theory is sometimes called the "internalist" model of technological development; R.J. Forbes termed it the theory of "self-generation", based on a presupposition that there exists an iron chain of causes and effects leading to the present Western milieu: he refused to countenance it. Other historians would join him in theory, but their actual works betray their acceptance of it in practice. Ortega y Gasset understood and criticised the attitude in the clearest possible terms:

> One of the purposes of the foregoing argument has been to warn against the spontaneous but injudicious tendency of our time to believe that basically no more than one technology exists, the present Euro-American technology, and that all others are but awkward stammerings, rudimentary attempts. I have opposed this tendency and embedded our present technology, as one among many others, in the vast and multiform panorama of human technology in its entirety, thereby relativizing its meaning and showing that every way and project of life has its corresponding specific form of technology.[27]

If the internalist theory continues to hold sway, the reason is not hard to find. Any internalist theory, implicit or explicit, has no logical answer to the real existence of other technological systems with other rationales and purposes. There are two elements here: the internalist theory must first attempt to minimize the influence of non-Western inventions on Western technological development. In the second place, if the internalist theory is to substantiate the presence of some internal dynamism or active principle in Western culture itself that explains its present technological status (as Bronowski attempted to do), it would then be hard put to explain the readily proven fact of alternative technologies that were invented by non-Western cultures, in the absence of Western culture and the elements in it that might have had something to do with technological development. Let me elaborate, briefly, on both issues.

Joseph Needham has something important to say about both. In the first place, he has turned out to be the single most important critic of the view that it is possible for scholars "to work backward from modern science and technology, tracing the evolution of scientific thought to the experiences and achievements of Mediterranean antiquity. An abundant literature exists in which we may read of the foundations laid by Greek and Roman thinkers, mathematicians, engineers and observers of Nature."

Another sinologist, A.C. Graham, pointed out that for two thousand years Greek rationality gave no technological advantage to those who had it over those who didn't. But this is probably not the most important issue. What *is* important is the number of inventions from India and China, to take just two nations, that once helped fill real gaps in the technological development of the West, and have now been sufficiently documented by Needham, Lynn White and others.[28] How important was the diffusion of these inventions to the West? A simple example, from the understanding of Bacon, would suffice:

> It is well to observe the force and virtue and consequences of discoveries. These are to be seen nowhere more conspicuously than in those three which were unknown to the ancients, and of which the origin, though recent, is obscure and inglorious; namely, printing, gunpowder, and the magnet. For these three have changed the whole face and state of things throughout the world, the first in literature, the second in warfare, the third in navigation; whence have followed innumerable changes; insomuch that no empire, no sect, no star, seems to have exerted greater power and influence in human affairs than these mechanical discoveries.[29]

All three "mechanical discoveries" were, of course, Chinese. Yet, Western scholars find it hard to acknowledge the Chinese origin. As Needham points out, even J.B. Bury, who in his *Idea of Progress* recognized the crucial role of these inventions, failed to point out, even in a footnote, that none of the three was of European origin.

To take up the second issue, I said earlier that if it could be shown that Western man has a special gift for technology, the nature of this ability would then be rooted in some dynamic principle of Western culture itself. The existence of other technological systems, however, would pose a real theoretical difficulty to such Toynbeean opinions. The easiest way to get out of this problem would be to completely ignore the worth of alternative technological developments. This, however, has had grave and unfortunate results.

Since entire generations, in Asia, Africa, and Latin America, have been taught to believe that they had no technical past or history to speak of. A piece of propaganda, which if taken to its rigorous conclusion, would imply that for the hundreds of years that these societies have survived (before Western contact) they have existed purely on sunshine or some form of manna that fell religiously from the heavens.

And if Western historians of technology have leaned over backward to explain why the West has "won" technologically, their colleagues in sociology, religion, psychology, anthropology, and history have found it "intriguing" to focus their analytical tools on non-Western civilizations to enquire into why these cultures did not germinate or nurture the kind of technological development evident in Western history recently. Religion, philosophy, tradition have been pulled out of context to be paraded as possible villains to technological progress in these cultures. From here, it has been but a step to look to the West for "transfers of technology".

Thus, our preoccupation with Western technology has resulted in more attention being paid by the new nations to transplanting elements of the Western technological system instead of updating their own indigenous ones. In the minds of planners at least, the latter will eventually be supplanted by "advanced" networks of production. A fruitless effort then to devote any serious attention to them.

Once, however, these categories of "advanced" and "pre-industrial" are seen to be the false distinctions they are, our attitude to technology in the world changes. If a technology is to be understood as *any* productive system that meets at least man's primary needs, then it would be foolish indeed to restrict our attention to solely the technology of the industrial nations, for large masses of people do not use that system. They use another and it also fulfil, its purpose, that of survival: witness the gradual increase of populations using indigenous means of subsistence production.

To use the terms introduced in this chapter, it is more than necessary that we begin to see that *homo faber* is not to be identified merely with Western *homo faber*: with the Western productive and cultural system. We can think too in terms of African *homo faber* or the Chinese or Indian editions of the model: but each on its own terms. Today, it is possible to see how the model has become identified with Western capacity for technology and culture. So we think that any technical problem can only be solved in the Western way. What indeed did the Southern nations and their peoples do to solve their technical problems before contact with the West in 1500?

They had their own production systems, of course, some of which are still in use. In part, then our study of the present could also be simultaneously a study of the past.

NON-WESTERN TECHNOLOGICAL SYSTEMS

In an article published in *Nature,* twenty-five years ago, Needham gently took a native scholar of Siam (Thailand) to task for disclaiming that his own people had failed to make any contribution to science. Needham then went on to mention the work of Loubere in the late seventeenth century on Thai science, indicating that the European had found sufficient grounds to be interested in what the Siamese knew. Needham went on to enquire about Siamese technology, particularly textile technology, in which the experience of long tradition had made possible a unique level of sophistication in the art.[30]

Twenty-five years ago, educated men in Sri Lanka were probably as alienated from their own technical histories as their Siamese counterparts. Today, the situation has changed for the better. A thorough study of the vast and ancient structures of hydraulic engineering has become the basis for a massive governmental effort to reactivate these in order to help solve Sri Lanka's irrigation problems.[31]

A solid contribution recently in print in Korea is Sang-woon Jeon's *Science and Technology in Korea*. The MIT historian of technology, Nathan Sivin, has put the work into perspective with a splendid foreword that needs quoting in some detail:

> Korea's science and technology are worth knowing and thinking about in connection with technology transfer for special reasons. Unlike China, Korea's styles in thinking systematically and objectively about nature and in developing instruments and techniques of material culture were always defined in the shadow of a large sophisticated nearby civilization. The Korean experience differs from Japan's in that its influences from China flowed in more freely and directly, across a shared land border or a short stretch of sea. It was from Korea in fact that new sciences and arts were carried into Japan during the early centuries until regular contact between Japan and China became possible. As recent Korean and Japanese scholarship begins to cohere, it is becoming plain that we have not yet adequately recognized what a great part immigrant Koreans played in the formative phases of Japanese civilization as men of learning, craftsmen, and indeed nobles. Korea thus presents for our reflection the case of a country seeking to maintain its identity against pressures too imminent to be shut out.[32]

Since this book is more interested in the issues surrounding technology in different societies, I will not go into any detailed description or summary of the Korean scholar's work: anyone wishing to have a proper picture will undoubtedly read the book on his or her own initiative and consult the extensive bibliography therein. The volume, however, is an important element in the war of those who seek to raise Korean confidence in the face of the massive, foreign technological system that haunts the borders of that land. Sivin captures the matter fairly well, I think:

(Jeon Sang-woon) is a Korean, and his pride in certain inventions and techniques is perceptibly greater than if he were a foreigner writing about Korean science. He knows that he is addressing a world-wide readership most of whom did not dream before they picked up his book that Korea is entitled to exert any claim upon the universal history of science. He knows that many educated people in Europe and the United States are just recovering from the shock of learning Joseph Needham's lesson, that the Chinese tradition is as indispensable as that of the early West in determining the potentialities of science. This book opens up still another range of awareness by demonstrating that peripheral societies must be examined with equal seriousness if we are not to overlook real originality. The author also knows that this implication will be equally surprising to most of his follow Koreans. In Korea today the power to exploit nature is seen as an importation, as foreign in its essence. Few people are aware that, say, Korea in 1400 may very well have had the most advanced astronomical-observatories in the world. Is it possible that science is not fundamentally Caucasian and Judeo-Christian (and all sorts of other things Koreans are not) after all?[33]

Science and Technology in Korea has an excellent chapter on astronomy, ancient observatories, sundials, and the measurement of time in general. Another entire chapter is devoted to meteorology; one more to physics and physical technology, and includes descriptions of printing, firearms, shipbuilding, civil engineering, and architecture. Two final chapters deal with chemistry, chemical technology and pharmaceutics, and cartography. Altogether, a preliminary investigation: the real history of science and technology in the service of Korean civilization is still to come.

THE CASE OF AFRICAN TECHNOLOGY

The case for *homo faber* in Africa is not so easily described, not because it does not exist, but because too many influential people have thought it does not exist. Though the myth of Africa, raised from the ground by Europeans, is now being gradually disestablished, it will be many years before Africa gets its own Needham.

As late as 1952, a former Governor of Kenya, Lord Milverton, could claim (and be believed) that the African had stagnated in primitive savagery during the period in which most other civilizations had been busy accumulating histories. The myth of African primitiveness was so widespread that whole generations of young Africans were brought up in the belief that Africa had no past.[34]

Today, this picture is undergoing a change. There are already eight good histories of African civilizations, based on new archaeological finds that have necessitated a more objective (and richer) image of the African continent: A Dutch geographer, Olfert Dapper, writing in 1668, for example, and describing the city of Benin, could only compare it with Amsterdam:

> The town seems to be very great. When you enter into it, you go into a great broad street, not paved, which seems to be seven or eight times broader than the Warmoes street in Amsterdam. . . .
>
> The king's palace is a collection of buildings which occupy as much space as the town of Harlem, and enclosed with walls. There are numerous apartments for the Prince's Ministers and fine galleries most of which are as big as those on the Exchange at Amsterdam. They are supported by wooden pillars encased with copper, where their victories are depicted, and which are carefully kept very clean.
>
> The town is composed of thirty main streets, very straight and 120 feet wide, apart from an infinity of small intersecting streets. The houses are close to one another, arranged in good order. These people are in no way inferior to the Dutch as regards cleanliness: they wash and scrub their houses so well that they are polished and shining like a looking-glass.[35]

Archaeology and contemporary accounts have resurrected the numerous civilizations of Nubia and Kush, Ethiopia and the Sudan, of the East Coast, the Bantus, of North Africa, the Sahara and the states of the West African forests, none of which could have arisen or flourished without an equally thriving technics, whether in agriculture or in industry.

On the other hand, it may also be true that Africa did not see any compelling need to build up a splendid technology as did the Chinese, for example, for with its vast reserves of gold and ivory it could acquire (very much like OPEC today) the commodities it needed through trade. As Jack Goody describes this trade:

> Africa was involved in vast networks of wide-ranging trade long before the Portuguese came on the scene. For East Africa we have a late first-century sailors' guide to the trade along the coast, the *Periplus of the Erythrean Sea*.

Long before the Europeans arrived there were trade routes from Madagascar up to the East African Coast, through the Red Sea and into the Mediterranean, along the Persian Gulf to India, South-east Asia and Indonesia. Possibly there was a direct route by which cinnamon was brought from the Spice Islands. By the time the Portugese reached the eastern shores of Africa, the Chinese had already been very active there; before the development of the gun-carrying sailing ship on the Atlantic seaboard, the maritime commerce of the Indian ocean made Western Europe seem like an underdeveloped area.[36]

As far as African technology is concerned, we will restrict ourselves to two main industries: metallurgy and textiles. The African iron-working processes have been detailed by Forbes,[37] but put into a larger perspective by Basil Davidson.[38]

The iron industry probably first reached significant proportions after the rise of Meroe, which has been described, perhaps with some exaggeration, as the "Birmingham of ancient Africa". By the middle of the first century B.C. smelting works on an extensive scale had already been initiated:

Sayce, who looked at Meroe some fifty years ago, could write that "mountains of iron slag enclose the city mounds on their northern and eastern sides, and excavation has brought to light the furnaces in which the iron was smelted and fashioned into tools and weapons". By the time of the building of Musawarat, in short, Meroe was the centre of the largest iron-smelting industry in Africa south of the Mediterranean coast.[39]

Davidson observes that a study of the African iron age is of key importance for an understanding of contemporary African origins, for,

Only with good iron tools would African peoples subdue the natural difficulties of living where they did, spread themselves across the land, flourish and multiply.[40]

So crucial was the role of iron that the Portuguese at the end of the fifteenth century found the king of the Congo a member of an exclusive "blacksmiths' guild" himself. And in the middle of the nineteenth century, in the area south of the Sahara, another foreigner would write:

the *enhad* (smith) is in much respect, and the confraternity is most numerous. An *enhad* is generally the prime minister of every little chief.[41]

From Kush, the technique of iron-making went further south, where we encounter another startling phenomenon: iron trade with India. Our source here is Edrisi, an Arab historian who visited the coast of East Africa in 1154. The people of the East coast, the Zanj, he wrote, own and work mines, trade in wrought iron and make large profits:

Hither come the people of the islands of Zanedji to buy iron and transport it to the mainland and the islands of India, where they sell it for a good price; for it is the object of a big trade there, and is in big demand.[42]

Edrisi went on to observe that the iron of Sofala (on the coast) was better than Indian iron, though he also added that "the Indians are masters in the arts of working it". We shall, later encounter this "Indian iron", which was in fact steel, produced and exported from India to Damascus for the manufacture of swords. Thus iron ore mined in south-eastern Africa, was forged in South-western India, fashioned in Persia and Arabia, to end up as the weapons and chain mail of the Saracens as they faced the Crusaders.

The art of making textiles forms round a few basic elements: materials and the knowledge of dyeing and printing. The citizens of Meroe wore silk from China and cotton fabric imported from India. Apart from these, they and most other African peoples were proficient in the manufacture of bark fibre cloth (very similar, in manufacturing process, to the making of paper). The resultant cloth could be watered, beaten, and joined up wherever necessary. The particular technique seems to have been perfected in Central Africa, in Uganda, and Tangannika, where craftsmen travelled from village to village meeting local needs. For resist, use was made of locally available materials, including mud and clay. The Yoruba people of the Niger region still practise *adire elek* dyeing, in which use is made of a paste made from cassava starch. The Soninke people of Senegal use rice paste for similar purposes.

The African is part of a culture that encourages the development of distinctive and sophisticated tastes: he will not accept or wear what he considers inferior fabric, crude design, or garish colours. Neither does he appreciate dull colours. With the result that he has forced industries outside Africa that serve his needs, to restructure their manufacturing processes to his cultural standards. Up to 1750, for example, textile goods from Manchester sent to Africa to displace Indian textiles, were sent back to England with the suggestion that the English industry either improve their imitation techniques or continue to provide genuine Indian originals.

Even today, industries in Great Britain, Holland, and Germany that continue to cater to the African market, submit meekly to the demands of African cultural taste. Stuart Robinson writes that European products are still produced with an indigo-type dye (Africans love indigo) and with a slight looseness of top colour (genuine indigo rubs off on the skin). African designs are simulated as faithfully as possible to produce the effects normally associated with hand-painted fabrics.[43]

Africa today is at the cross-roads[44] seeking an identity that is able to take up and continue, in refinement, the experience of its valuable past. Technology, African man's ability to use his hand and brain, has played a significant role in the shaping of this past. New technology, where it is in keeping with the abilities of African *homo faber,* and in so far as African governments use it to re-establish indigenous techniques and processes in which there is popular technical confidence, should be increasingly Africanized, as has happened in the case of those industries outside Africa catering to the African textile market. The cultural re-awakening of Africa is the strongest among the Southern nations today. But it cannot of itself do much, unless it is accompanied by an acceptance of African man's technical ability to solve his own problems, so long denied in opinion.

Our final foray in this chapter is into the large continent of Latin America, including Mexico. Not much time, indeed, need be spent on the technologies of the Maya, the Aztecs, and the Incas: the documentation is still continuing and new archaeological discoveries are constantly being announced.[45] There is, for example, an entire volume devoted to the textile processes of Peru.[46] Everywhere the comment made is that the pre-Columbians accomplished such a variety of works using such a restricted set of tools.

In trying to assess the quality of these civilizations, Fernand Braudel was not very certain that they could be placed among the higher civilizations of the time. He wrote:

Do the Aztec or Mexican civilizations and the Inca or Peruvian civilization – have full right to be placed at that level? The answer is yes as far as ability, brilliance, art and original turn of mind are concerned. It is equally so if we consider the ancient Mayas' wonderful science of calculation and the longevity of these civilizations: they survived the terrible impact of the Spanish Conquest. On the other hand, the answer is no when we note that they used only hoes and digging sticks; that they had no large domestic animals (except llamas, alpacas and vicunas); that they had no knowledge of the wheel, arch, cart or metallurgy in iron.[47]

The assessment, I submit, is unfair. In the first place, as Benjamin Lee Whorf, who studied the languages of these civilizations noted, the predilections of one culture are not necessarily those of another, nor do they need be so. In technology, it is imprecise to criticize the lack of metallurgy when these peoples had access to large deposits of obsidian which they had learnt to work into weapons and tools. Nobody criticizes the United States of America for using wood for 90 per cent of its energy needs up to 1850 and even later, even when the use of coal had already been well established by English technicians. And Lewis Mumford has exploited the fact of Maya and Inca roads as independent proof "that broad streets and even highways are not a mere by-product of wheeled chariots or carriages. Religious processions and military parades both have need for them."[48]

The vandalism of the conquistadores who melted down the Indians' priceless and exquisitely fashioned gold and silver works of art to convert them into money, is repeated by scholars on the level of ideas, when the latter submit non-Western phenomena to the criteria of their own. When the Latin American, however, enters the National Museum in Mexico, he can see immediately what he was once capable of:

miracles of artistic creation expressed in the carvings, pottery, and textiles of contemporary and bygone ages, and a traditional lore which ranged from the sophisticated abstractions of the Mayan calendar to a wealth of legend and popular poetry.[49]

Once, without the benefit of the European, he was expert in the art of terracing fields, paving roads, and constructing suspension bridges high up in the Andes.

Perhaps, something may be said of the social organization of these peoples, for which the Incas, for example, showed such genius, so much so it has even been described as "socialist" or "communist" by different scholars. In fact, a young *mestizo,* Jose Carlos Mariategui (1895-1930), once suggested that the traditional social groups of the Indians, the *ayllu,* should be maintained and strengthened in the face of further aggrandizement by large landowners. Mariategui suggested that the Indian's strong instinct for communal action should be taken into account in working out the future transformation of his country, which he at any rate thought to be inevitable. And he was able to defend the *ayllu* "not on abstract principles of justice nor for sentimental traditional reasons, but on sound practical and economic grounds".

Tanzania has recognized something similar in its recent scheme concerning *Ujamaa* villages and Chinese commune-developments confirm the issue: that there are other forms of viable economic activity than plain rugged Western individualism. Part of the land reform movement in Mexico depended a great deal on the re-establishment of *ejidos,* cooperative landholdings similar in some respects to the Peruvian *ayllu.* As Stephen Clissold summed it all up:

> "The Latin America of today is the China of Yesterday; the China of Today is the Latin America of Tomorrow" is a "poem" which has been given recent currency in Cuba. But most thinking Latin Americans have no wish to see their countries as another China, another Russia, another North America, or another Europe. What they wish to see is a Latin America which is truly itself, which has explored and harmonized its own diverse potentialities, rediscovered its past, and incorporated the still living structure into its personality, a Latin America which looks to others only in order to be genuinely itself. A Latin America which has, in the fullest sense of the term, acquired a mind of its own.[50]

This, I think, is how it always should have been and should continue to be. As Western cultural history is but a strand in a number that hold together a larger geography of the human mind, there is great violence perpetrated in attempting to examine the whole, through structures associated too closely with the part.

Volumes have been written on the evolution of a new world culture, most of them unacceptable. For, a world culture could only evolve on a basis of equal participation by all existing cultures. The present cultural domination of the West could otherwise merely lead to a world culture not very much different from the Western particular and thus inevitably serving Western interests, ideals, and goals. Joseph Needham has already pointed out, for example, that the philosophical bases of most international organizations are Western, and ultimately have through the years benefited the further extension of the power of the Western world.[51]

Before we think in terms of applying one cultural pattern to the face of a global civilization, it would be more in the interests of a "democracy" of cultures if we first made available to ourselves all the valleys, have given". The same holds true for technological systems, unless we wish to extend the sway of a gravely-faulted megamachine over every nook and corner of the world.

It is not my intention here to repeat what others, including Mumford, Roszak, Ellul, Dubos, and Reich, have written about the Western technological system and the problems for man, animal, and nature built into its very structure. But with Nathan Sivin I think it possible to argue throughout this book, and convincingly, that the model of social and technological development idealized out of the industrial revolution in England, the United States, and certain parts of Western Europe is no longer the sole means by which the Southern countries and nations of Asia, Africa, and Latin America can hope to survive.

CHAPTER TWO

Indian Technology and Culture: 1498-1757

> The sciences and technologies of the non-European world had different seekings and developments to those of Europe. Further, in countries like India, their organization was in tune with their more decentralist politics and there was no seeking to make their tools and work-places unnecessarily gigantic and grandiose. Smallness and simplicity of construction, as of the iron and steel furnaces or of the drill-ploughs, was in fact due to social and political maturity as well as arising from understanding of the principles involved. Instead of being crude the processes and tools of eighteenth century India appear to have developed from a great deal of sophistication in theory and an acute sense of the aesthetic.
>
> —Dharampal: 1971

I am afraid, this chapter, because of its subject and of the nationality of the person writing it, may sound as ideological as some of the volumes of the Soviet encyclopaedia. And I have also emphasized Indian technology at the expense of the other facets of Indian life. Yet, so great a quantity of paper and print has been devoted to Indian philosophy and art, and so pervasive is the opinion abroad that these aspects of the Indian mind have remained saturated with "spirituality" and "world-denying" tendencies, that it has seemed but natural to conclude that technology or material culture could not have been attended to in the measure desired.

The idea of India's other-worldliness was systematically emphasized by Max Muller, though he was not the first.[1] The West, in fact, came to its acquaintance with the Indian mind through the translations of the Indian scriptures first published by the French scholar, Duperron, in 1802. It was through these translations that Schopenhauer read the Upanishads and came to call them the solace of his life and death.

What turned out to be a solace for Schopenhauer has proved to be a menace to those interested in interpreting the realities of India's economic and technological experience. A number of works have over the years been devoted to the theme that sees India's poverty as somehow related to this supposed predilection for other-worldliness.[2] As late as 1972, an Indian writer, Subhayu Dasgupta, set out to argue in his book, *Hindu Ethos and the Challenge of Change,* that the decisive obstacle to India's economic development was the "stagnant Hindu mentality", fortified and maintained by the caste system.[3]

An improved version for the popular audience, but parading a similar theme, is available with Alan Watts:

> A king of ancient India, oppressed by the roughness of the earth upon soft human feet, proposed that his whole territory should be carpeted with skins. However, one of his wise men pointed out that the same result could be achieved far more simply by taking a single skin and cutting off small pieces to bind beneath the feet. These were the first sandals.
>
> To a Hindu, the point of this story is not its obvious illustration of technical ingenuity. It is a parable of two different attitudes to the world, attitudes which correspond approximately to those of the progressive and traditional types of culture. Only in this case the more technically skilful solution represents the traditional culture, in which it is felt that it is easier for man to adapt himself to Nature than to adapt Nature to himself. This is why science and technology, as we know them, did not arise in Asia.[4]

There is more clever nonsense in this than truth. It remains to be seen whether Watts and others of his ilk have not been unduly influenced by a view of the Indian mind that arose during a specific period of historical writing. The Indian scholar, Pratap Chandra, poses and partly answers these questions:

> Our historical judgement has been coloured by this ardent desire to wish away all heterogeneity from ancient India. In the ideological sphere, it has resulted in the conjuring up of an intellectual monolith supposedly governing the Hindu ethos for centuries. The creators of this stereotype were frankly biased in favour of absolutism and spiritualism, and they made sure that Indian thought became predominantly, if not exclusively, a collection of idealistic and absolutistic views.
>
> Moreover, their preoccupation with Western thought and its categories and terminology easily persuaded them to view Indian thought analogously with it. Consequently, terms like "orthodox" and "heterodox", "established viewpoint" and "heresy" crept into all the accounts of Indian thought without anyone's asking whether these terms were relevant in the Indian context. . . . Indian thought in this way became fixed in the popular mind as an instance of a unilinear stream-like growth in which absolutism formed the "main

current" and other views became either its imperfect expressions or deviations from it.[5]

This myth of the absolutist nature of Indian philosophy was repeated so often everybody came to accept it as gospel truth: the average Hindu was turned into a dreamy visionary and his philosophy into a world-and-life-denying dogma. Abroad, thinkers wrote eulogies to the fine quality of Indian philosophical speculations, and Indians themselves exhilarated for a long while over this world-wide apprecia- tion and praise. Yet, when Max Weber's analysis appeared and seemed to indicate that it might have been precisely Indian religion that had impeded economic development, a new crisis of confidence arose. Weber had set out, wrongly in my view, to "inquire as to the manner in which Indian religion, as one factor among many, may have prevented capitalistic development (in the occidental sense)."[6]

To conclude this preliminary issue, there is still no evidence to date that one line of thought was so dominant that it had become "orthodox": there was no single established view that set itself up against and above a number of other supposed heresies. The hypothesis that Hindu philosophy (wrongly identified wholly with the idealistic school) proved disastrous for India's technical capacity or economic activity should be relegated to the garbage heap; so too, the fashionable cliche that India spent too much of its energy in the pursuit of metaphysical speculations. Those who still want to hold on to the older myths are advised to read carefully what follows.

INDIAN AGRICULTURE

Before industrialization revolutionized England and implicitly devalued agriculture, most of the civilizations of the world functioned as plant economies.[7] Agriculture constituted the basis of all other human activities. Indian agriculture, practised for about four thousand years, made possible a civilization so complex and varied that some elements of it remain incomprehensible even today. Yet, the art of agriculture was exercised within the conscious limitations seen in the ecology of the land.

Dr. Wallick, a Superintendent of the East India Company's Botanical Garden at Calcutta, was heard by the English Commons' Committee on this issue, on the 13th of August, 1832. We might add, he was also one of the first to talk in terms of technology being "inappropriate":

The husbandry of Bengal has in a great measure been misunderstood by the Europeans out of India. The Bengali husbandry, although in many respects extremely simple and primeval in its mode-and form, yet is not quite so low as people generally suppose it to be, and I have often found that very sudden improvements in them have never led to any good results. I have known, for instance of European iron ploughs introduced into Bengal with a view of superseding the extremely tedious and superficial turning of the ground by a common Bengali plough. But what has been the result? That the soil which is extremely superficial, as I took the liberty of mentioning before, which was intended to be torn up, has generally received the admixture of the under soil, which has deteriorated it very much.[8]

He was asked whether the techniques could be improved. He answered in the negative:

Certainly, but not to so great an extent as is generally imagined; for instance, the rice cultivation. I should think, if we were to live for another thousand years, we should hardly see any improvement in that branch of cultivation.[9]

Twelve years earlier (1820), Colonel Alexander Walker had prepared a more comprehensive report on the agriculture of Malabar and Gujarat. We quote extensively, since the document will not be widely available for quite some time. The entire report may be found in Dharampal's *Indian Science and Technology in the Eighteenth Century,*[10] a volume that contains a fine selection of documents, prepared by English visitors to India during the colonial period. Comparatively speaking, the book as a whole is as important for a study of Indian technology as is Sang-woon Jeon's volume for the Korean. Wrote Walker:

In Malabar the knowledge of Husbandry seems as ancient as their History. It is the favourite employment of the inhabitants. It is endeared to them by their mode of life, and the property which they possess in the soil. It is a theme for their writers; it is a subject on which they delight to converse, and with which all ranks profess to be acquainted.

Their sacred Bulls, and their superstitious regard for the cow, have their foundation in the great service they render to Husbandry. Under all these circumstances of favour and encouragement, we should expect that it would be the study of this people to improve the art of cultivating the ground, and that they would in such a length of time have discovered the most convenient and effectual instruments for the purpose. This however has been strenuously denied by those who wish to accommodate the ideas and habits of European Husbandry to that of Malabar. They reproach the Hindus for employing rude and imperfect instruments. This censure cannot apply equally to every part of India where various instruments are in use, and of different construction. The plough is the first and foremost important machine in agriculture. In Gujarat it is a light and neat instrument. It has no Coulter but has a sheathing of iron: the furrows of the Husbandman are as straight as a line, and of sufficient depth to produce the most abundant crops. This is the real and only useful test of good farming.[11]

Walker observed that the Malabar plough was light enough to be carried on a man's back; more important, it was accommodated to the soil, which was light, unobstructed by stones and softened with water.

In a climate where the productive powers are so great, it is only necessary to put the seed a little way into the ground. . . . It must be a strong proof that the Indian plough is not ill adapted for its purpose, when we see arising out of the furrows it cuts, the most abundant and luxurious crops. What can be desired more than this? The labour and expense beyond this point must be superfluous. The Indian peasant is commonly well enough informed as to his interest, and he is generally intelligent and reflecting. He is attached to his own modes, because they are easy and useful; but furnish him with instruction and means, and he will adopt them, provided they be for his profit. He will not be led away by speculation and theory, which he cannot afford to follow; but he will not refuse any more economical, and less laborious mode of cultivation.[12]

Walker then goes on to recall an experiment done at Salcette, in Goa, where European ploughs and implements were delivered to natives to use in their fields. The experiment failed: the plough was too heavy, the oxen more easily fatigued, the implements more expensive. Most important, the whole package failed to produce any results better than those earned with the traditional plough. As usual, Walker notes, the failure was imputed to the prejudice, sloth, and obstinacy of the villagers. Walker then proceeded to make two conclusions, which are still valid today:

Before we charge them with ignorance and obstinacy for neglecting to adopt our recommendations, we should first be sure of two things; that the new system would give them more abundant harvests, at less expense and labour; and that we have taken all the means and care that were within our power, for their instruction in the art. It should also be well considered how far our agricultural process is suited to the cultivation of rice, the great crop of India, and of which we have no experience. We ought also to remember that India has very little occasion for the introduction of new plants for food. There are more kinds of grain cultivated perhaps than in any other part of the world.... I am at a loss to know what essential present we can make to India. She has all the grains that we have and many kinds more of her own.[13]

Walker then presents a description of the drill plough, the system of rice transplantation, the variety of agricultural implements, "some of which have only been introduced into England in the course of our recent improvements". He notes the different kinds of ploughs adapted to the differing soils and seeds, the mallets, harrows, and rakes, the methods of hoeing and weeding. Presenting the care the Indian farmer devoted to the performance of his art, he writes:

The numerous ploughings of the Hindoo Husbandman have been urged as a proof of the imperfection of his instrument; but in reality they are a proof of the perfection of his art. It is not only to extirpate weeds that the Indian Husbandman re-ploughs and cross-ploughs; it is also to loosen the soil, apt to become hard and dry under a tropical sun; and hence it becomes necessary to open the earth for air, dew and rain. These advantages can only be obtained by exposing a new surface from time to time to the atmosphere. In India dews fall much more copiously than they do with us, and they are powerful agents in fertilizing the soil.[14]

He has also something to say about ecological farming, the cultivation of different crops within the same field, each beneficial to the other; he mentions the culture of fodder for cattle. "It would require a volume to pursue all the details of Hindoo Husbandry." The fencing and enclosure of fields; the broad grassy margins for pasture. The whole world, he exclaims, "does not produce finer and more beautifully cultivated fields than those in Gujarat".

He acknowledges a debt to Col. Wilk's *History,* where he has seen a note, based on the result of observation and study on the spot. "It shows that the practice of the Indian farmer is founded on the most enlightened principles of modern farming." There are more details about the cultivation of fruit trees, the collection of manure, the rotation of crops.

In every part however that I have visited the application of manure for recruiting and restoring land is well understood. The people seem to have all the resources that we have in this respect.[15]

He next produces an extract from a letter of a friend, "whose intelligence and opportunities of observing the practice of Indian

Husbandry, are not I believe exceeded by any man in that country".

In Gujarat, and indeed in the Deccan, but specially in Gujarat, careful and skilful agriculture is probably as much studied as in England. In many points an English farmer might condemn the practice at first sight; but in time he would learn that much of what he did not approve, under an idea that the same system in all respects that succeeds in England ought to be followed here, was of the first importance, was in fact what constituted the great means of success in this climate and that to depart from the existing practice would be folly.[16]

Walker's document contains too many details to summarize here *in toto*: there is a more detailed account, for instance, of the agriculture of Malabar itself, of the fifty kinds of rice grown in the region, of the knowledge the local population has of the kinds of soils and how they should be treated. "The system of the natives is too well founded on experience to reject."

A conclusion echoed by Dr. Voelcker, a Consulting Chemist with the Royal Agricultural Society of England in 1889, when he was deputed to make inquiries and suggest improvements to Indian agriculture. Relevant extracts:

On one point there can be no question, viz. that the ideas generally entertained in England, and often given expression to even in India, that Indian agriculture is, as a whole, primitive and backward, and that little has been done to try and remedy it, are altogether erroneous. . . . At his best the Indian Ryot, or cultivator is quite as good as, and in some respects the superior of, the average British farmer; whilst at his worst, it can only be said that this state is brought about largely by an absence of facilities for improvement which is probably unequalled in any other country and that the Ryot will struggle on patiently and uncomplainingly in the face of difficulties in a way that no one else would.

Nor need our British farmers be surprised at what I say, for it must be remembered that the natives of India were cultivators of wheat centuries before those in England were. It is not likely, therefore, that their *practice* should be capable of much improvement. What does, however, prevent them from growing larger crops is the limited facilities to which they have access, such as the supply of water and manure. But to take ordinary acts of husbandry, nowhere would one find better instances of keeping land scrupulously clean from weeds, of ingenuity in device of water-raising appliances, of knowledge of soils and their capabilities, as well as the exact time to sow and to reap, as one would in Indian agriculture, and this not at its best alone, but at its ordinary level. It is wonderful, too, how much is known of rotation, the system of mixed crops and of fallowing. Certain it is that I, at least, have never seen a more perfect picture of careful cultivation, combined with hard labour, perseverance and fertility of resource, than I have seen in any of the halting-places in my tour.[17]

All these accounts belong to the nineteenth century, but numerous descriptions of an earlier period reinforce the later picture. Abul Fazl, for example, found agriculture flourishing "in high degree" in Bihar, where rice "which for its quality and quantity was rarely to be equalled".[18]

The variety of agricultural produce is well documented too. Writing about the indigenous plantations of south India, Buchanan noted the practice of having a separate piece of ground allotted for each kind of plant. "Thus one plot is entirely filled with rose-trees, another with pomegranates, and so forth." The coconut tree supplied a great deal of necessities: pith, liquor, fruit, "cloths", roofs, sails, and ropes. In Bengal, notes another traveller, "the plantations have no end". He mentioned mangoes, oranges, citrons, lemons, pineapples, coconuts, palm-fruits, and jack-fruits. Stavorinus adds bananas and guavas. Other fruits, grown in large-scale plantations, included melons, apples, peaches, figs, and grapes. Ives refers to "the endless variety of vegetables" used by Indians in their curries and soups.

Bengal itself produced a surplus that was traded all over the country: grains, spices, and pulses. "To mention all the particular species of goods that this rich country produces is far beyond my skill." Rice was grown in such plenty that, writes Orme, "it is often sold at the rate of two pounds for a farthing". According to Dow, much of the land in Bengal had turned into desert through oppressions and famines (1770-71). Even so, Bengal continued to produce "for double the number", that is, for thirty million people. There are further reports about the agriculture of Bihar, Rajputana, Rohilkhand, Malwa Flat, Gujarat, Khandesh, and the west coast as a whole.

In general, the valleys of all rivers consisted of "one sheet of the richest cultivation". Berar, with its black soil, produced cotton, wheat, barley, and flax. Nagpur wheat matured in three months. The Northern Circars are described as "the granary of the Carnatic". The spices of Malabar, including pepper, ginger, cardamom, and cinnamon found their way into Europe.[19]

Of *irrigation technology,* too, there is a great deal to be said, though it might be added at once that the Indians probably had no great need of the kind of massive hydraulic works with which the Chinese engaged themselves.[20] The opinion, however, that India's irrigation works were of little or no consequence has been so influential that even Indian historians have glibly accepted. R.C. Majumdar, for example, is quite certain of "the comparative absence of artificial irrigation" in eighteenth-century India.[21] Contrast this with Alexander Walker's comment that,

> the practice of watering and irrigation is not peculiar to the husbandry of India, but it has probably been carried there to a greater extent and more laborious ingenuity displayed in it than in any other country.[22]

In Bengal, dykes were the usual response to floods, and tanks and reservoirs stored water if rains proved scarce. Wells were a common feature; even today, every village continues to have its own well. Where there were no rivers, deep extensive tanks, measuring from three to four hundred feet at their sides, were constructed, with a short temple alongside for adornment. Ghulam Hussain writes about Bengal:

> Rivers, small and large, are plenty in this country and the practice of digging tanks is very common. People in this country seldom drink the water of wells because everywhere the water of rivers and tanks is found in abundance. And generally the water of the wells is salt, but with a little digging of the soil water comes out.[23]

Elphinstone reports that expensive embankments had been constructed on the rivers of Khandesh for irrigation purposes, and in Rohilkhand the local chiefs had built aqueducts "traversing corn-fields in all directions". In the hilly regions, dams blocked streams. Bishop Heber, in the early part of the nineteenth century described Bharatpur State as "one of the best cultivated and watered tracts which I have seen in India". To conclude with Walker:

> The vast and numerous tanks, reservoirs, and artificial lakes as well as dams of solid masonry in rivers which they constructed for the purpose of fertilizing their fields, show the extreme solicitude which they had to secure this object.
>
> Besides the great reservoirs for water, the country is covered with numerous wells which are employed for watering the fields. The water is raised by a wheel either by men or by bullocks, and it is afterwards conveyed by little canals which diverged on all sides, so as to convey a sufficient quantity of moisture to the roots of the most distant plants.[24]

INDIAN INDUSTRY

Next to agriculture, cotton and cotton goods constituted the principal industry in the Indian sub-continent, as did the woollen industry in England. Up to 1800, no country produced a greater abundance or variety of textiles in the world. China remained the only close rival. In 1700 itself, India was the largest exporter of textiles in the world. Wrote Dubois:

> With such simple tools the patient Hindu, thanks to his industry, can produce specimens of work which are often not to be distinguished from those imported at great expense from foreign countries.[25]

A world today that cannot understand production except in terms of high energy inputs, complex machines, and processes and massive organization, merely to imagine how fine manufactures could be effected in such large quantities through the simplest tools is a difficult task; but one after another of the foreign travellers in India remarked on the "perfection of the manufacture" and of the simplicity and imperfection of the tools used.

The loom provided the basis of the Indian industry, particularly in the eighteenth century. It provided employment to "hundreds of thousands of inhabitants, comprising the weaver caste" and to "countless widows" and families, who engaged themselves in the subsidiary processes of cotton spinning. The weaving industry itself was extensive, stretching from "the banks of the Ganges to the Cape". "On the coast of Coromandel and in the province of Bengal, when at some distance from the high road or a principal town, it is difficult to find a village in which every man, woman or child is not employed in making a piece of cloth."

The fact is the textile industry was highly coordinated with agriculture. Indeed, it was usually during their "vacations from agriculture", that is, when the crops were growing or had just been harvested, that one found a great number of villagers applying themselves to the loom, "so that more silk and cotton manufactured in Bengal than in thrice the same extent of country throughout the empire and consequently at much cheaper rates".

In the north, the great Moghuls maintained *kharkhanas* (factories) for their specific needs. Elsewhere, native princes preserved their own arrangements. And one economist has noted how this constant source of employment declined and withered,[26] as the princes fell prey to the machinations of British power.

From Roman times till their decline in the nineteenth century, the main textile areas on the sub-continent had been the same:

They are described in the *Periplus* of the first century A.D. in much the same terms as they were described by travellers of the seventeenth and eighteenth centuries. These main areas were three: *Western India,* with Gujarat, Sind, and Rajputana as the focus; *South India, comprising the Coromandal Coast as it used to be known, stretching from the Kistna Delta to Point Calimere; and North-east India* including Bengal, Orissa and the Ganges Valley.[27]

Each of these areas specialized in specific classes of fabrics and even employed techniques indigenous to the region itself, with different designs, motifs, and symbols. Here I shall discuss merely the first, for reasons of space.

From Abbe de Guyon, in the middle of the eighteenth century, we have the following account of Ahmedabad in western India:

People of all nations, and all kinds of mercantile goods throughout Asia are to be found at Ahmedabad. Brocades of gold and silver, carpets with flowers of gold, though not so good as the Persian velvet, satins, and taffetas of all colours, stuffs of silk, linen and cotton and calicoes, are all manufactured here.[28]

Surat, "an emporium of foreign commerce", manufactured the "finest Indian brocades, the richest silk stuffs of all kinds, calicoes and muslins".

Painted and printed calicoes constituted the most important class of Indian fabric exported from Surat in the seventeenth century. They covered a wide range of quality, the best and the more expensive being *painted* rather than *printed*. . . . In the former case, dyes and mordants were applied to the cloth, not with a wood-block, but free-hand with brush. Thus, each painted design had the character of individual drawing with the human and sensuous touch, instead of being limited to the repeat pattern imposed by the print-block. Sometimes painting and printing techniques were combined, but the finest decorative calicoes from both western India and the Coromandel Coast were of the painted kind.[29]

There has been among textile historians a controversy whether Indian craftsman *printed* their chintz with mordants, the original assumption being that this technique (like the drill plough in agriculture) had first been discovered in Europe. In 1966, however, the controversy was effectively settled in India's favour, when the Roques manuscript was discovered in the archives of the Bibliotheque Nationale in Paris: the manuscript contained a detailed account of the textile industry and manufacturing processes, including textile *printing, observed by the writer in western India.*[30]

The chief centres of cotton painting in western India were Sironj in Rajputana and Burhanpur in Khandesh. Cheap printed cottons came from Ahmedabad, though these were also produced in the regions devoted to the painting technique. Gujarat also produced embroideries on quilts and coverlets, but this industry saw decline by 1690, when the centres of European trade had shifted to Sind and the Punjab. The century also saw the development of the carpet industry, almost certainly due to the emperor Akbar (1556-1605) who is known to have encouraged the immigration of Persian craftsmen for the purpose. The industry was located principally at Agra and Lahore.

The woollen industry was situated in Kashmir, which produced the extraordinary cashmere shawls, whose beauty was considerably "enhanced by the introduction of flower work". The wool was imported from Tibet, after which it was bleached and manufactured. As for silks, in western India, fabrics from them were often mixed with cotton. True silks were worked as *Patolas* in Patan, Gujarat. Printed silk, *culgar,* is still produced in the same places today as in those times, in the form of saris of artificial, printed silk or *kalgers*. One species of cotton and silk fabrics consisted of *alachas,* striped fabrics, later consciously imitated in England. The *cuttanee* was a satin weave; the cheapest of the mixed fabrics were called *tepseils*, produced for the West African trade. And for the Portuguese demand, there were silk and wool fabrics, called *camboolees,* produced in Sind.

The calicoes themselves ranged from the finest and the most expensive muslins, *seribaffs,* wanted in the Islamic countries, to the cheaper varieties of coarse cloth, *dungarees* and *gunny*. Intermediate calicoes, *baftas,* were woven at Broach, Nosari, and Surat. There were calicoes patterned on the loom, woven from different coloured threads, and other calicoes dyed after weaving. The colours remained bright after washing, and this proved to be their great attraction.

I have presented this varied, brief, picture of the textile industry, and added a great many vernacular names, not because I wish to bring some local colour to this book, but merely because I want to emphasize the width and breadth of the Indian influence in the period. Irwin and Schwartz give the following interesting list of the

Company's orders in the west coast for a single year (1695-96) as a sample:

20,000 Pallampores large
10,000 Pallampores midling
10,000 Pallampores small
2,000 Quilts large, new patterns, $3\frac{1}{4} \times 3$ yds.
2,000 Quilts midling
5,000 Quilts small
10,000 Chints Culme 2
20,000 Chints Caddy3, as much variety of works and stripes as may be
10,000 Chints broad, 9 × 1 yd of variety of new patterns
20,000 Chints narrow
10,000 Serunge (Sironj) Chints, the best and newest works and paintings on good strong cloth, $\frac{1}{2}$ stripes, $\frac{1}{2}$ flowers
20,000 Chints Paunch Runge 4
5,000 Chints Surat[31]

The south Indian textile industry was concentrated on the Coromandel Coast and was specificially directed to the spice trade in the Malay Archipelago. The economies of the Spice Islands had no need for bullion or any other foreign commodity: Indian cloth formed the only acceptable means of exchange. The trade was tri-cornered. Arabs carried bullion to the Coromandel Coast, exchanged these for textiles, exchanged the latter in the islands for spices, and returned with these to the Middle East.

The various exports of the third area for textile production, Bengal, literally boggle the mind: they included ordinary silk, mixed silk and cotton goods, calicoes, linens, mulmuls, tanjebs, chintzes, ginghams, pure silk and woollen fabrics, turbans and shawls. Unwrought silk from Kasimbazar itself totalled 300,000 to 400,000 pounds weight, and went to form the base for manufactures in Europe. Of the country around Kasimbazar, Grose was able to claim that the workers there "generally furnish 22,000 bales of silk a year, each bale weighing a hundred pounds". This constituted a boom, which it was, and towards the end of the seventeenth century, most Europeans found these silks cheaper than the French and Italian silks they had till then patronized.

Within another fifty years, this entire picture would be to a great deal reversed. In England and the Continent, the textile industries were being revolutionized through the study and close imitation of the work of Asian craftsmen. And later, these improvements, harnessed to the machine, would turn the tide of events.

INDIA TEACHES EUROPE

The second volume of the Singer *History of Technology* concludes with a curious and startling opinion, which reads:

> (But) this volume has also seen to its end a relationship between east and west that we shall not encounter again. When the Middle Ages closed, the east had almost ceased to give techniques and ideas to the west and ever since has been receiving them.[32]

I shall show that this is bunk, that Singer, a medical scholar, did not know that British surgeons learnt the art of plastic surgery from Indian practitioners, to take just a random example. But it is the history of the development of the textile industry in the West *after* the Middle Ages that absolutely refutes his claim. There is no doubt that particularly after 1500, India's influence was crucial to the European textile industry. Country after country in Europe tells the same tale.

I am not talking here of trade, which normally supplied the genuine and costly Oriental fabrics that were coveted by the European nobility. I mean the vast *imitation* industry that sprang up to cater to the commoners, who, unable to afford the expensive originals, had to be content with copies, and this not merely in England, but also in France, Germany, the Netherlands, Spain, Switzerland, and America.

Thus, P.R. Schwartz and R. de Micheauiex, in their book, *A Century of French Fabrics: 1850-1950,* state that in France "the term *indiennes* (chintz) is found in Marseilles inventories since at least 1580, and on 22 June, 1648, a card-maker and engraver of this too was associated with the dyeing of cloth to make *indiennes*".[33] The imitation printing of these chints was banned in due course, but the *indiennes* continued to grow in popularity, "despite the heavier fines imposed, the ripping off by the police of the offending print dresses from the backs of women walking in the streets and the destroying of stocks of garments".[34] Once the ban was lifted (1759), the designers began to introduce designs at first based solely upon Oriental patterns.

The same may be observed of Germany, where in order to protect the home industry, Frederick William I banned the wearing, importing, or selling of any kind of printed or painted calicoes. Again, these laws were flouted and in 1743, print works were established in various parts of the country; imitation printing being officially permitted by 1752.

Textile workers in Italy, from the late seventeenth century to about 1855 had their earlier patterns based on *indiennes*. More obvious is the case of the Netherlands:

The Dutch merchants and explorers were some of the first to bring back the painted and printed Coromandel cloths from the East during the early seventeenth century... and Dutch textile printers attempted to imitate the brilliantly coloured Indian cottons which were not only fast to water but became more beautiful and brilliant when washed. Their first attempts with the oil or water colours long used in Europe, that either smelt badly or would not wash, bore no comparison with the Eastern cloths printed or painted with mordant dyes and indigo.

The first European print works was founded in Amersfoot in Holland in 1678 and attempted to use Indian methods.[35]

Success came after nearly seventy years, when Dutch printers succeeded in copying the sheer Indian cottons by using copper plates.

The first Spanish calico print works started by the Esteban Canals in Barcelona in 1738, copied *indiennes* and used the imported Eastern textiles as a source of pattern. Switzerland repeats the story, and in the United States too, the earliest evidence of textile printing shows Eastern influences in the patterns.

It has not been any different with the circulation of ideas in Europe. Literature-wise, three large documents found in European libraries are representative, having been written with the express purpose of informing Europeans about Indian processes and techniques. The letters of the Jesuit, Coeurdoux, for example, were sent out in 1742 and 1747. The earlier letter begins typically:

I have not forgotten that in several of your letters you have urged me to acquaint you with the discoveries I might make in this part of India, since you are persuaded that knowledge is to be acquired here which, if transmitted to Europe, would possibly contribute to the progress of science or to the perfection of art. I should have followed your advice sooner, had not almost continuous occupation taken up all my time. Recently, with a little leisure, I have used it to find out the way in which Indians make these beautiful cloths, which form part of the trade of those Companies established to extend commerce, and which, crossing the widest seas, come from the ends of Europe into these distant climes to search for such things.[36]

The second letter begins in a similar vein, though it is actually an introduction to a letter written to Coeurdoux by a certain M. Lepoivre, on Indian processes, to which Coeurdoux has added a lengthy comment.[37] It is the sequel to these letters, however, that is interesting:

Father Coeurdoux's actual descriptions contain many further processes in the printing and the dye preparation and these were studied and commented upon by Edward Bancroft (1744-1821), the distinguished English chemist, in his book, *Experimental researches concerning the philosophy of permanent colours,* London, 1794 and 1813. It was also used in G.P. Baker's outstanding work, *Calico painting and printing in the East Indies in the 17th and 18th centuries,* London, 1921, and in a number of Continental books and journals of the seventeenth and eighteenth centuries.[38]

The second manuscript, called the *Beaulieu Manuscript,* discovered also quite recently, dates about 1734. It was quoted in the treatise on cotton painting by the Basle manufacturer, Jean Ryhiner (1728-90), which was first written in 1766, though not published till 1865. It was also discussed in the book written by Chevalier de Querelles, *Traite sur les toiles Peintes,* Paris, 1760. It still exists, together with eleven actual samples of painted cloth brought back by de Beaulieu, as well as full details of the processes discovered.

The third manuscript, also recently come to light, is the Roques manuscript, of 333 pages, and in which, as I noted earlier, the priority of wood-block printing in India is firmly established.[39]These documents are representative; there are many more, like the anonymous article to be found in the *Journal economique,* Paris, July 1752. It begins thus:

> There can be no doubt but that it would be most harmful to the State were we to neglect our own production of light silken and woollen materials in favour of Persian or Indian cottons. It can, however, only be a good thing to know how these peoples set about applying colours to their cotton cloths, which not only do not run or fade when washed but emerge more beautiful than before. Everyone can see for himself how useful this would be when he envisages what the possibilities could be for our cotton, linen and hemp cloth.[40]

The writer then goes on to describe two Indian methods for the dyeing process.

All this knowledge, of course, crossed the seas free of charge. Father Coeurdoux milked the neophytes he had recently baptised for his information. And the painter, Pieter Coeck Van Aelst (father-in-law of Pieter Brueghel the Elder), transported back to Holland the techniques of indigo resist painting after his travels in the East in the early sixteenth century. Yet, once this knowledge and expertise entered the European countries, it soon made its way into the gradually evolving patent system. As early as 1676, in England, William Sherwin, an engraver of West Ham, took out

> a grant for fourteen years of the invention of a new and speedy way for producing broad calico, which being the only true way of the East India printing and stayning such kind of goods.[41]

The history of iron and steel in India has not thus far seen consistent review, neither will I attempt one here, for it would take a volume. I shall rest content with a few illustrations at best, and indicate the influence of Indian steel on the industrial revolution in England.

As Dharampal observes, there are a number of accounts concerning the production of iron and steel in India during the Vasco da Gama epoch; in fact, one of the earliest ones, numbering about seven pages, is to be found in the book of D. Havart, a Dutchman, entitled, *The Rise and Decline of Coromandel,* originally issued in Utrecht in 1692. Dharampal's volume itself contains three accounts dated 1795, 1829, and 1842. Here I shall say something about steel, about whose development in India there is too little in the public imagination. Then, everyone called it not steel, but *wootz.*

The Celtic smiths of Noricum, a Roman province (today, Lower Austria), made good steel as early as 500 B.C. and traded it to Italy. There were other centres during the Iron Age where steel was produced by holding wrought iron in the charcoal of the forge until it reached white heat and then quenching it, but the resulting product did not reach Celtic standards. The latter itself, however, was not as good as the so-called Damascus steel, the only true spring steel known before the Age of Gunpowder. And this steel was made in India, as early as the 5th or 6th centuries B.C. in the Hyderabad district by smiths through a process of fusion known as *wootz.*[42]

By the 1790s, a sample of wootz had landed in England, where it roused considerable scientific and technical interest. It was examined by several experts, found in general to match the best steel then available in England, and, as one observer put it, "promises to be of importance to the manufacturers" of Britain. He also found it "excellently adapted for the purpose of fine cutlery, and particularly for all edge instruments used for surgical purposes".[43] Demand increased, so that 18 years later, one frequent user could write:

I have at this time a liberal supply of wootz, and I intend to use it for many purposes. If a better steel is offered to me, I will gladly attend to it; but the steel of India is decidedly the best I have yet met with.[44]

The man who felt Indian steel as being of some importance to the manufactures of Britain was none other than Stodart, the person who later assisted Faraday in preparing and investigating a large number of steel alloys. According to Hyene, Stodart was "an eminent instrument-maker", and according to another writer, a man named Pearson who was assisted by Stodart in conducting the experiments on wootz in 1794-95, the latter was "an ingenious artist".[45]

Nineteenth-century England produced very little of its steel from its own iron. In one year alone (1823) it imported more than 12,000 tons of iron, mostly from Sweden, to work into steel. The quality of iron ore in the country was decidedly low for the purpose, and so was the fuel. The English, on examining Indian wootz, applied their own experience to conclude "that it is made directly from the ore; and consequently that it has never been in the state of wrought iron". Dharampal writes:

Its qualities were thus ascribed to the quality of the ore from which it came and these qualities were considered to have little to do with the techniques and processes employed by the Indian manufacturers. In fact it was felt that the various cakes of wootz were of uneven texture and the cause of such imperfection and defects was thought to lie in the crudeness of the techniques employed.

It was only some three decades later that this view was revised. An earlier revision in fact, even when confronted with contrary evidence as was made available by other observers of the Indian techniques and processes, was an intellectual impossibility. "That iron could be converted into cast steel by fusing it in a closed vessel in contact with carbon" was yet to be discovered, and it was only in 1825 that a British manufacturer "took out a patent for converting iron into steel by exposing it to the action of carburetted hydrogen gas in a closed vessel, at a very high temperature, by which means the process of conversion is completed in a few hours, while by the old method, it was the work of from 14 to 20 days.[46]

The founder of the Indian Iron and Steel Company, later also connected with Sheffield, J.M. Heath, soon discovered that the wootz process combined both the British discoveries mentioned above. But he went on to add that while the Indian method lasted two hours and a half, the processes at Sheffield required four, "to melt blistered steel in wind furnaces of the best construction, although the crucibles in which the steel is melted are at a white heat when the metal is put into them, and in the Indian process, the crucibles are put into the furnace quite cold".

He concluded, of course, by denying that the Indian producer had any theory of his operations, since the process was discovered by scientific induction, and the theory of it only explained in the light of modern chemistry. Such a conclusion is strange, but it was easy during the ethnocentric period he lived in to propose it. Even the British Royal Society betrayed the spirit of the times: a letter describing wootz as having a "harder temper than anything we are acquainted with" was altered to "anything known in that part of India".[47]

INDIAN MEDICINE

The literature on Indian medicine is enormous, rich and various, and I do not again intend to paraphrase it all here.[48] In what follows, I shall merely describe two of the more important medical arts of India, practised particularly through the time or period under study: plastic surgery and inoculations against small-pox. Both were indigenously evolved and the accounts we have, come from Westerners sent out to study them. In the case of plastic surgery, the world of medicine's debt to India is easily acknowledged in every volume on the subject. The second, inoculation, may come as a surprise, though not to those conversant with similar discoveries about Chinese medicine.

Colonel Kyd, an Englishman, had the following to say regarding general surgical skill in the subcontinent:

> (In) chirurgery (in which they are considered by us the least advanced) they often succeed, in removing ulcers and cutaneous irruptions of the worst kind, which have baffled the skill of our surgeons, by the process of inducing inflammation and by means directly opposite to ours, and which they have probably long been in possession.[49]

Dr. H. Scott wrote the following letter on January 12, 1792, again on the subject of Indian surgery:

> In medicine I shall not be able to praise their science very much. It is one of those arts which is too delicate in its nature to bear war and oppression and the revolutions of governments. The effects of surgical operation are more obvious, more easily acquired and lost by no means so readily. Here I should have much to praise. They practise with great success the operation of depressing the chrystalline lens when become opake and from time

immemorial they have cut for the stone at the same place which they now do in Europe. These are curious facts and I believe unknown before to us.[50]

One of these curious facts was the inoculation against small-pox disease, practised in both north and south India till it was banned or disrupted by the English authorities in 1802-3. The ban was pronounced on "humanitarian" grounds by the Superintendent General of Vaccine (following Dr. Jenner's discovery in 1798). There are two detailed accounts of the Indian method in Dharampal's volume.

Small-pox has a long history in India: it is discussed in the Hindu scriptures and even has a goddess devoted exclusively to its cause. It seems therefore almost natural to expect an Indian medical response to the disease. The inoculation treatment against it was carried out by a particular tribe of Brahmins from the different medical colleges in the area. These Brahmins circulated in the villages in groups of three or four to perform their task.

The person to be inoculated was obliged to follow a certain dietary regime; he had particularly to abstain from fish, milk, and *ghee* (a form of butter), which, it was held, aggravated the fever that resulted after the treatment. The method the Brahmins followed is similar to the one followed in our own time in certain respects. They punctured the space between the elbow and the wrist with a sharp instrument and then proceeded to introduce into the abrasion "variolus matter", prepared from inoculated pistules from the preceding year. The purpose was to induce the disease itself, albeit in a mild form: after it left the body, the person was rendered immune to small-pox for life.

The Brahmins had a theory of their operations. They believed the atmosphere abounded with *imperceptible animalculae* (refined to bacteria within a larger context today). They distinguished two types of these: those harmful and those not so. According to Dr. J.Z. Holwell, FRS, who addressed the President and members of the College of Physicians in London concerning this:

That these animalculae touch and adhere to everything, in greater or lesser proportions, according to the nature of the surfaces which they encounter; that they pass and repass in and out of the bodies of all animals in the act of respiration, without injury to themselves, or the bodies they pass through; that such is not the case with those that are taken in with the food, which, by mastication, and the digestive faculties of the stomach and intestines, are crushed and assimilated with the chyle, and conveyed into the blood, where, in a certain time, their malignant juices excite a fermentation. . . which ends in an eruption on the skin.

They lay it down as a *principle,* that the *immediate* (or instant) cause of the small-pox exists in the mortal part of every human and animal form; that the *mediate* (or second) *acting* cause, which stirs up the *first*, and throws it into a state of fermentation, is multitudes of *imperceptible animalculae* floating in the atmosphere; that these are the cause of all epidemical diseases, but more particularly of the small-pox.[51]

The Brahmins therefore believed that their treatment in inoculating the person expelled the immediate cause of the disease. How effective was the inoculation? According to Holwell:

When the before recited treatment of the inoculation is strictly followed, it is next to a miracle to hear, that one in a million fails of receiving the infection, or of one that miscarries under it.[52]

A later estimate by the Superintendent General of Vaccine in 1804 noted that fatalities among the inoculated counted one in 200 among the Indian population and one in 60 or 70 among the Europeans. There is an explanation for this divergence. Most of the Europeans objected to the inoculation on theological grounds. But, also significant, the social context of Indian medicine was being changed.

I have said that the inoculation of the Brahmins induced the disease, although in a mild form. The risk inherent in such a treatment becomes obvious: the disease might spread by contagion from those inoculated *with* it to those not so treated. Certainly, this was not the problem when the operation was universally practised and everyone underwent it. This universality ceased to obtain with the arrival of the British. Like many specialists in India, including teachers, the Brahmin doctors had been maintained through public revenues. With British rule, this fiscal system was disrupted and the inoculators left to fend for themselves.

In such a situation even vaccine inoculation could find little acceptance. In 1870, another Superintendent General of Vaccine wrote that the people were still reluctant to get vaccinated because of the general opinion that the indigenous inoculation possessed "more protective power than is possessed by vaccination in a more damp climate". Thus, the indigenous method still continued to win some allegiance. For the areas round Calcutta in 1870, it was estimated that only ten per cent of the population had not been so inoculated; for Bengal, the figure was 36 per cent.

The experience of plastic surgery, happily, followed more fruitful channels.[53] The art of plastic surgery or rhinoplasty rose again as a perfect response to a peculiar Indian custom: the cutting off or amputation of the nose as a punishment for crime or as a plain humiliation. The resulting disfigurement drove the sufferer to a class of surgeons who soon founded a thriving business in the reconstruction of noses. In 1794, Dr. H. Scott would refer to the "putting on noses on those who lost them" and send to London a quantity of *caute,* the cement used for "uniting animal parts".

The earliest of these rhinoplasties were performed in India already in 1600 B.C. and there are still families that practise the same method today. The operation is described in the *Sushruta Samhita,* a book written in 600 B.C. by the well-known Indian surgeon, Sushruta: a flap from the cheek was cut off to reconstruct the nose. Later, a better method used flaps from the forehead. Another Indian surgeon, Vagbhat, provided a more detailed description in his book, the *Ashtanga Hridyans,* in the fourth century A.D. Twice did this art spread from India to the rest of the world. Writes S.C. Almast.

In the centuries which followed the golden age of Ayurveda, the knowledge of rhinoplastic procedures was probably transferred to Western civilization by the free interchange of thought and experience between Hindu, Arab, Persian, Greek, Nestorian and Jewish scholars. Celsus the Roman who lived 25 B.C. to 50 A.D. was probably the earliest Western European to describe plastic operations on the nose. The *Sushruta Samhita* was mentioned as *Kitabe Sushrud* by Ibn Abi "Usaybia" (1203-1269 A.D.), the first historian of Arabian medicine, in his book. It was also stated that during the reign of Al-Mansur (died 775 A.D.) an Indian medical work by Sushrud was rendered into Arabic by Manke, the Hindu court physician by the suggestion of Wazie Yahyaibn-Khalid.

The practical secret of rhinoplastic operations spread from India through Arabia and Persia to Egypt and from there it leaked to Italy. In the 15th century in Sicily, Branca used cheek flaps to reconstruct the proud noses of hot-blooded swordsmen. His son Antonio tried flaps from the arm and by the late 16th century Tagliacozzi had published his work on the Italian method of arm flap rhinoplasty.[54]

It was only two centuries after that, in the nineteenth, that German, French, and English surgeons could study the entire method afresh, through the translation of the Sanskrit literature and personal observations through travel in India:

In Kumar near Poona a Mahratta surgeon was seen by two medical officers of the East India Company, James Findlay and Thomas Cruso, performing a rhinoplasty by the median forehead flap. This case was reported as a "singular operation" in the *Madras Gazette* of 1793. The patient was Cowasjee, a Mahratta bullock driver with the British army in the war of 1792. He was taken prisoner by Tipu Sultan who cut off his nose and one of his hands. He went back and rejoined the Bombay army of the East India Company and after one year had his nose reconstructed in Kumar near Poona. A description of this case also appeared in the *Gentleman's Magazine* of London in a letter from India in 1794.[55] The description of the "singular operation" was responsible for the later spread of this technique to European countries and to the United States of America. The first successful case of forehead flap rhinoplasty performed in England was published in 1814, about twenty years after the Cowasjee case. Carpues' book *An account of Two Successful Operations for Restoring a Lost Nose from Integument of the Forehead* was published in the year 1816 and helped to create a considerable interest in this subject. In Germany Carl Ferdinand Von Graefe performed the first total reconstruction of the nose in 1816 and coined the term "plastic surgery" in his text on this subject published two years later. Jonathan Mason Warren from America undertook rhinoplasty by the Indian method in the year 1834. Captain Smith published his *Notes on surgical cases – Rhinoplasty* in the *British Medical Journal* in 1897 and suggested improvements. Keegan (1900) wrote a review of rhinoplastic operations describing recent improvements in the Indian method.[56]

If plastic surgery began in India with the reconstruction of noses, the Europeans took over the principles underlying the method: not living in a culture that insisted on visiting punishments with the lopping off of noses, they could more easily see a possible application to other areas of bodily defect. A similar case of development of techniques *after* diffusion concerns the *noria*, another peculiar Indian invention, originally thought to have originated in Egypt, but now generally acknowledged as Indian.[57]

It is perhaps easier now to understand the opinion of the traveller, Robert Orme, who, though often critical, was forced to admit that "the arts which furnish the conveniences of life have been carried by the Indians to a pitch far beyond which is necessary to supply the wants of a climate which knows so few". He went on to remark paradoxically that though the knowledge of the Indians in "mechanical matters is very limited", Europeans were "left to admire, without being able to account for" the manner in which, for example, the people built their bridges and constructed their huge temples.

Or their ships, for that matter. In the middle of the eighteenth century, John Grose noted that at Surat the Indian ship-building industry was very well established, indeed. "They built incomparably the best ships in the world for duration", and of all sizes with a capacity of over a thousand tons. Their design appeared to him to be "a bit clumsy" but their durability soundly impressed him. They lasted "for a century". Lord Grenville mentions, in this connection, a ship built at Surat which continued to navigate up to the Red Sea from 1702 when it is first mentioned in Dutch letters as "the old ship" up to the year 1770.[58]

Grenville also noted that ships of war and merchandise "not exceeding 500 tons" were being built "with facility, convenience and cheapness" at the ports of Coringa and Narsapore. The Parsees in Bombay were known as great builders of ships – and skilled as "naval architects". In *Les Hindous,* Solvyns, after introducing about 40 sketches of boats and river vessels used in the Indian north in the 1790s, observed that "the English, attentive to everything which related to naval architecture, have borrowed from the Hindoos many improvements which they have adapted with success to their own shipping".[59] Needham too sees the multiple masts of India and Indonesia in this light.

Dr. H. Scott sent samples of *dammer* to London, as this vegetable substance was used by the Indians to line the bottom of their ships; he thought it would be a good substitute "in this country for the materials which are brought from the northern nations for our navy. . . . There can be no doubt that you would find *dammer* in this way an excellent substitute for pitch and tar and for many purposes much superior to them."

I have left out of this review a number of other technical processes used by the Indians before and during the colonial period, including the making of paper, ice, armaments, the breeding of animals, horticultural techniques, and so on. A tiny note now on Indian science.

Needham wrote in 1963:

> We cannot forejudge what the future developments of the history of science will bring forth, but if India was probably less original than China in the engineering and physico-chemical sciences, Indian culture in all probability excelled in systematic thought about Nature (as for example in the Samkhya atomic theories of *Kshana, bhutadi, paramanu,* etc.), including also biological speculations. . . . When the balance comes to be made up, it will be found I believe, that Indian scientific history holds as many brilliant surprises as those which have emerged from the recent study of China – whether in

mathematics, chemistry, or biology, and especially the theories which were framed about them.[60]

Indian scholars have been too generous, and Western scholars too niggardly about the nature of these "brilliant surprises", and much of what Indian scientists really produced has been clogged by controversy. But things are changing; recently, even a diehard, one-track mind as Rostow's showed signs of a more enlarged education in this regard. All I can do right now is to indicate the literature available in the field, and let the reader pursue the issue himself or herself.[61]

THE INDIAN MIND

My brief illustration of Indian science and technology should not be construed as constituting the total interpretation of the Indian *homo faber* paradigm. For the technology of India can be related to other aspects of an Indian philosophy, or, an Indian mind. Take textiles.

The colours used by the Indian craftsmen for their textiles, writes Robinson, were not only brilliant and of great variety, they were also often exceedingly subtle and particularly so in their tonal qualities:

> Their colours seem to contain hidden qualities and effects that only appear in differing lights. The *pagris* or headwear produced in Rajasthan (originally Rajputana), Kotah and Alwar contained two slightly differing shades which produced a constantly changing colour pattern as the fabric rippled.[62]

It may surprise the reader, but a similar "moving" principle underlines the painted frescoes of the Ajanta caves. In describing these frescoes, Richard Lannoy, the most scintillating interpreter of Indian culture to date, writes:

> At first sight it is the genial "Buddhist humanism" which strikes the visitor [to the fresco caves]. Yet these reassuringly human scenes are not quite what they seem to be. For one thing, even the best preserved are exceedingly elusive to "read"; one must make an appreciable effort to slow down one's reading of their visual language in order to perceive the spatial and tactile relations established between the figures. There is no *recession* – all *advance* towards the eye, looming from a strange undifferentiated source to wrap around the viewer.
>
> This is not an optical illusion of cave-light; on close examination it will be found to result from a controlled use of almost equal *tones* in the variation of local colour. A patch of green, say, juxtaposed to a patch of red, is of very nearly the same tonality when photographed in monochrome. Because of this

tonal equality one is constantly discovering new figures which were unseen through the deliberately unaccented or "suppressed" tonality of detail, and the tempo of this slow discovery is very precisely calculated. Every figure has a counterfigure, every body an anti-body. Each figure is inseparable from its environment. The optical basis of this technique is very simple and is frequently used by Bonnard, Vuillard, and Matisse to obtain a hallucinating, visionary effect; the later, psychedelic poster artists made a trick of it. One can assume that the Ajanta painters discovered the effect under similar lighting conditions. There is one vital difference, however; at Ajanta there is no source of light in the caves, a fact which says much about the metaphysic of the cave sanctuaries. Objects are their own light when experienced by *all* the senses in harmony, and such harmony was the goal of the cave ritual.

When viewed by flickering light, as was intended, only fragmentary glimpses of the colours and lines of the objects depicted can be obtained. A body undulates towards the eye from an indistinguishable blur; moments (perhaps minutes) later, a second body wells out of the blur and is seen to be intertwined with the first. The viewer is so involved in this optical assimilation that his relation to the other figure only proceeds gradually from the tactile to the emotional recognition of its significance. It cannot be reduced to verbal interpretation, as it is pure tactile sensation.[63]

Lannoy himself admits that his interpretation of these frescoes came to him in the company of C.V. Raman, who also won the Nobel Prize for Physics. He also acknowledges that the experience made him change his mind that there is any sort of antithesis between the unified aesthetic sensibility nurtured by Indian culture through the ages and the demands of the scientific method.

He then went on to make a sharp distinction between the "single, fixed viewpoint to which we (in the West) are conditioned by the artifice of optical perspective" and the "multiple-perspective, shifting viewpoint employed in the portrayal of figures, animals and objects at Ajanta".

Here, I may be allowed to deviate a little to indicate one more example of the imperialism of categories that has led to a grossly distorted view of the abilities of non-Western artistic personalities. In most art history volumes, the appearance of perspective has often been seen as a form of "development" in the regular refinement of techniques in art: this criterion has then been universalized without the recognition that another culture may refuse the technique, not out of ignorance, but out of choice: this is confirmed by the insistent indifference of the Indian artist to the single pyramidal tableau contained within a border, which is the commonest structure for the Western-type image. In *The Speaking Tree,* Lannoy not only distinguishes Indian art from Western, he also emphasizes its distinctness from Chinese art, a point well made after the general confusions spread abroad by Northrop.[64]

The entire metaphysic of the Ajanta frescoes is too complex to get across here in brief. More fruitful, perhaps, is a minor discussion on Indian architecture, which again may be sharply distinguished from its Western counterpart.

Lannoy has pointed out that the caves of India are the most singular fact about Indian art, and he is right, for they serve to distinguish it from that of other civilizations. A prodigious amount of labour, spread over a period of about 1,300 years, was expended in this "art of mass", the excavations of rock sanctuaries and monasteries. These caves were hewn out of solid rock; in other words, they were "constructed" through the excavation of space.

These sanctuaries were cut from nearly-perpendicular cliffs to a depth of a hundred feet: in all cases, this excavation was carried out by means of a chisel 3/4 inches wide; the same chisel was also used to carve out elaborately decorated columns, galleries, and shrines. The two largest structures of the kind are staggering in their dimensions:

> The Kailash temple at Ellora, a complete *sunken* Brahmanical temple carved out in the late seventh and eighth centuries A.D. is over 100 feet high, the largest structure in India to survive from ancient times, larger than the Parthenon. This representation of Shiva's mountain home, Mount Kailash in the Himalaya, took more than a century to carve, and three million cubic feet of stone were removed before it was completed. An inscription records the exclamation of the last architect on looking at his work: "Wonderful! O How could I ever have done it?"[65]

In Europe's middle ages, the great cathedrals, including the one of Chartres, rose from the ground upwards to the sky, supported not so much by stone as by the powerful religious symbolism that drove the Christian church. In India, the craftsmen did not build, but removed the earth and stone to discover space in the service of a different religious symbolism, not one identified with any religious monolith, but instead, one to which different religious groups owed allegiance. Lannoy is more precise:

> A hollowed-out space in living rock is a totally different environment from a building constructed of quarried stone. The human organism responds in each case with a different kind of empathy. *Buildings* are fashioned in sequence by a series of uniformly repeatable elements, segment by segment,

from a foundation *upwards* to the conjunction of walls and roof; the occupant empathizes with a *visible* tension between gravity and soaring tensile strength. Entering a great building is to experience an almost imperceptible tensing in the skeletal muscles in response to constructional tension. *Caves,* on the other hand, are scooped out by a *downward* plunge of the chisel from ceiling to floor in the direction of gravity; the occupant empathizes with an *invisible* but sensed resistance, an unrelenting pressure in the rock enveloping him; sculpted images and glowing pigments on the skin of the rock well forth from the deeps. To enter an Indian cave sanctuary is to experience a relaxation of physical tension in response to the implacable weight and density of the solid rock.[66]

Such an analysis of an Indian strategy to the experience of the Life-World could be reinforced by further investigations into Indian literature forms, music structures, and language. The similarities between the worlds of Sanskrit drama and the Ajanta caves, and the strucural disaffinities between Kalidasa's *Shakuntala* and Shakespeare's *King Lear* will not hold us here, but something might indeed be said about Indian music, for this art (with Indian dance) is one of the few living and thriving expressions of Indian cultural life today.

The Indian system of *talas*, the rhythmical time-scale of Indian classical music, has been shown (by contemporary analytical methods) to possess an extreme mathematical complexity. The basis of the system is not conventional arithmetic, however, but more akin to what is known today as *pattern recognition*. To quote Lannoy again:

In the hands of a virtuoso the *talas* are played at a speed so fast that the audience cannot possibly have time to *count* the intervals; due to the speed at which they are played, the *talas* are registered in the brain as a cluster configuration, a complex Gestalt involving all the senses at once. While the structure of the *talas* can be laboriously reduced to a mathematical sequence, the effect is subjective and emotional. . . . The audience at a recital of Indian classical music becomes physically engrossed by the agile patterns and counter-patterns, responding with unfailing and instinctive kinaesthetic accuracy to the terminal beat in each *tala*.[67]

This ability with instruments is repeated with the voice. The extraordinary degree of control of the human voice has been described by the musicologist, Alain Danielou, who has stated that Indian musicians can produce and differentiate between minute intervals (exact to a hundredth of a comma, according to identical measurements recorded by Danielou at monthly recording sessions). This sensitivity to microtones is, from the purely musicological point of view, of little importance, like the mathematical complexity of the *talas*. Nevertheless, as Lannoy puts it:

> It is an indication of the care with which the "culture of sound" is developed, for Hindus still believe that such precision in the *repetition* of exact intervals, over and over again, permits sounds to act upon the internal personality, transform sensibility, way of thinking, state of soul, and even moral character.[68]

CHAPTER THREE

Chinese Technology and Culture: 1368-1842

> It is logical to look at the Chinese view of China, because they are Chinese and it is their own country and their own history. It is necessary to make what is perhaps a special effort to see the Chinese point of view, because on the whole we have not paid enough attention to it in the past. Indeed there have been many occasions over the last century and a half of regular contact between China and the West when the prevailing attitude towards China revealed more about ourselves than about the Chinese.
>
> – John Gittings: 1973

Early this century, A.H. Smith put forth the opinion that Chinese history "is remote, monotonous, obscure and worst of all, there is too much of it". How much of this history, reconstructed principally by Western scholars, is useful for our purposes here is a moot question. That it has not been wholly useful for the Chinese themselves is evident from the pages of *China Reconstructs,* a journal that has busied itself with a re-interpretation of the pattern of the Chinese past that is more in keeping with the new consciousness of the Chinese Revolution, or, as W.F. Wertheim put it, with the Chinese belief in a secular promised land, *a better earth.*

Thus, what the Chinese feel about their past, particularly about their technology and their culture, would be difficult to identify with what Western scholars have felt about the same Chinese past. There is no question here about Chinese scholars being more objective than their Western counterparts, or *vice versa*. Bodo Wiethoff wrote, in his recent volume *An Introduction to Chinese History,* that "European conceptions of China have rarely reflected Chinese reality, but have first and foremost been the response to European needs".

The twists and turns that the Chinese image underwent in the West have all been well documented in the numerous representative histories of the period,[1] and need not concern us here: the point has been made. From the wide-eyed wonder of Marco Polo and the other medieval travellers, to the Jesuits' acknowledgement of the sophistication of the culture they faced, to Herder's impression of an "embalmed mummy, painted with hieroglyphs and swaddled in silk", Engel's "decaying semi-culture at the end of the earth" and Karl Marx's scorn concerning the "hereditary stupidity" of the Chinese, is indeed a long journey, but it sketches indeed the insecure personality of Europe more faithfully than it ever did the character of China itself.

Even Chinese scholars, educated in Western universities, have not been able to refrain from manipulating Chinese history to reinforce conclusions reached earlier by means frankly ideological, though at least *their* activities are intelligible in the light of the almost total devaluation of the role of the intellectual in the life of modern China. Thus, the late Lin Yutang, born and raised in China, but with his spirit moulded in the United States, was ready to confess (in 1937) a complete lack of confidence in the regenerative powers of his own people:

> Today China is undeniably the most incoherent and chaotic nation on earth, the most dramatically weak and impotent, the most incapable of rising up and marching ahead. . . .[2]

And that father and son team, Ch'u Chai and Winberg Chai, writing for an American audience, found it worth while to extol the Chinese past in *The Changing Society of China,* so they could find it easier to discuss the Mao government with a great deal of dismay and an unmistakable pinch of hate.

It was not that there were no scholars sympathetic to the Chinese cause in those times. In 1958, W.F. Wertheim sharply noted:

> The failure of many intellectuals in the West to understand fully the implications of the revolution in China may be symptomatic of a certain rigidity occurring in Western culture which may damage its adaptability to new developments in the world.[3]

Wertheim, and Jan Romein, both tried to express at that early date a premonition that the countries of Latin America, Africa, and Asia were inexorably moving into a position from which they would feel compelled, in their own interests, to see and write history on their own terms. The importance of China for us here lies not in any use it might have for the Chinese, however, but in that it gives us a very clear instance of independence from the West, from the Western interpretation of Chinese history.

The Chinese history I shall discuss now is of a very restricted period: that which began with the dynasty of the MING (1368/1644), followed later by the dynasty of the CH'ING (1644/1911) or the foreigner Manchus. Even with the latter, I shall restrict the description to 1842, the date of the Unequal Treaties, and the beginning of Western political dominance.

THE GREAT NON-QUESTION OF CHINESE HISTORY

It is precisely the interpretation of this specific period of Chinese history that has caused the greatest controversy, much of which has to do with the scientific and technological development of China. The celebrated question asks why China did not produce either modern science or an industrial revolution on the English pattern, especially since Chinese technology at least had reached a level of sophistication not yet attained in any other part of the world, as late as the fifteenth century.

I am aware that the discussion which will follow now might more suitably have been taken up in the earlier chapter on Indian technology and science. However, the fact that most of the debate round this all-important issue has been taking place in the Chinese context, and even with the active involvement of that great Sinologist, Dr. Joseph Needham himself, has dictated its being considered here. The reader, however, is advised to retain the qualification that the conclusions reached here are equally applicable to other contexts, including those presented by the histories of Islamic and Indian science. With that said, we can now proceed to set out first a few important distinctions.

It is necessary to make a very clear distinction between what are known as the scientific, industrial, and technological revolutions. Learned scholars, including Sinologists like A.C. Graham and Needham, tend to identify all three, which is incorrect. Stephen Toulmin and June Goodfield have been two of too few scholars to notice the differences:

> For, however spectacular the influence of science on a few branches of technology, one can easily exaggerate its impact on industry at large. Hero of Alexandria made hydraulic toys for his patrons, Galileo made calculations

about the strength of beams, but the age of applied science proper began only after A.D. 1850. In the interaction between theory and practice, science has again and again been in the position of debtor, drawing on the craft tradition and profiting from its experience rather than teaching craftsmen anything new. It has been said that "science owes more to the steam-engine than the steam-engine owes to science", and the same thing is true more generally. In its early stages, especially, the craft tradition was – so far as we can tell – devoid of anything which we would recognize as scientific speculation.[4]

In other words, though the scientific revolution (the discovery of how to discover) preceded the industrial revolution, it did not contribute to it as much as is normally assumed. The industrial revolution, in fact, was an empirical revolution, evolving in large measure independently of the scientific revolution. On the other hand, both these revolutions should in turn be distinguished from the technological revolution, often called the second industrial revolution, when science turned to improve practice and revolutionized technics. Toulmin and Goodfield note:

Before A.D. 1850, intellectual advances within the sciences of matter no more led to immediate improvements in the crafts than had Newton's theory of planetary motion at once led to better planetary forecasting.[5]

Two other historians of science, Derek J. de Solla Price and Rupert Hall, have written in similar terms.[6] "The time-lag in each instance between the establishment of a new craft-skill," observed Hall, "and the effective appearance of scientific interest in it, is of the order of 250 years, and in each of these examples it appears *after* the scientific revolution was well under way."

And now, to another distinction.

Not merely Needham, but a number of other writers have made a critical distinction between traditional science and universal science or Galilean science. In fact, all civilizations have attempted science, if by that word is understood any systematic abstract thought about nature. Nathan Sivin puts it the following way:

"Abstract" may have the sense not only of defining concepts on a more general plane than that of concrete sensual experience, but also that of seeking objective driving forces of change within nature itself rather than, like religion and magic, looking for explanations in terms of conscious will or emotion.[7]

This is the reason I felt it worth while to talk about Indian *science* in the earlier chapter, though few would wish to believe that it exploited the Galilean method. More recently, the Islamic scholar, Seyyed Hossein Nasr, has made a similar point in his splendid volume on Islamic science.[8] As he prefaces his work:

Islamic science, which is taken in this work to include disciplines concerned with the study of the cosmos, embraces a wide spectrum of intellectual activity, from the study of plants to algebra, carried out over more than a millennium by many races and peoples spread over the middle belt of the earth from Spain and Morocco to eastern Asia. *Because of its traditional character, this science is not limited in scope or meaning as is the modern discipline with the same name.* The Islamic sciences, even in the more limited sense considered here, which exclude the religious and many branches of the philosophical sciences, are considered at once with the world of nature, of the psyche and of mathematics. Because of their symbolic quality, they are also intimately related to metaphysics, gnosis and art, and because of their practical import they touch upon the social and economic life of the community and the Divine Law which governs Islamic society.[9]

In a review of the fourth volume of Needham's *Science and Civilization in China,* Lynn White remarked that Chinese science is of interest for those interested in the interpretation of East Asian cultures. Nathan Sivin, thinking along similar lines, has asked whether the label "anthropological" might not express what we have in mind:

If we use the first definition of science as systematic abstract thought – which might be called "anthropological", since it lets the Chinese theoretical encounter with nature define its own boundaries – the structure of knowledge looks very different indeed from that of modern science. It becomes as individual, in fact, as the map of scientific thought in ancient Greece or medieval Europe.[10]

Sivin goes on to observe, in line with the basic themes of this book that each culture breaks up its experience of much the same physical world into manageable segments in very distinct ways, which is very close to Benjamin Whorf's conclusions on the connections between languages and cosmologies. In each culture in fact certain basic concepts are consolidated in use due to their general usefulness in making nature comprehensible:

In Europe after Aristotle's time among the most important of these notions were the Four Elements of Empedocles and the qualitative idea of a proper place that was part of the definition of each thing. In ancient China the most common tools of abstract thought were the yin-yang and Five Phases concepts, implying as they did a dynamic harmony compounded out of the cyclical alternation of complementary energies. Today scientists use a wider range of well-defined concepts, embracing space, time, mass energy, and information.

Thus the fields of science in a given culture are determined by the application of these general concepts, suitably refined, re-interpreted if necessary, and supplemented by more special concepts, to various fields of experience, demarked as the culture chooses for intrinsic and extrinsic reasons to demark them.[11]

In the light of the new basic concepts available to modern science, traditional or "anthropological" Islamic, Indian, European, or Chinese science could be labelled "proto-science". The point to remember is that without these "proto-sciences" the foundations of Galilean science might never have been possible in the first place, as no one can see the latter as having had some form of spontaneous generation.

It is at this point that we enter controversy: scholars in the history of science have tended to re-examine the past of other cultures to try and discover why these did not make the breakthrough to universal science. The issue is then reduced to a rather misleading debate concerning the peculiar social or philosophical factors that *could* or *might have* inhibited the rise of, say, Chinese science. The hidden assumption being that only within the Western tradition itself there were conditions sufficient enough to permit the rise of universal science. As Graham notes, this has become increasingly hard to prove. Sivin puts it thus:

The basic concepts that [the Chinese] used to explore physical phenomena are precisely ying-yang, the Five Phases (wu hsing), the trigram and hexagram systems. . ., and others that used to be invoked (even by twentieth-century Chinese thinkers) as chiefly responsible for the failure of Chinese to learn how to think scientifically.

But to place the responsibility there is to commit one of the most elementary fallacies of historical explanation, namely to present a description of what the world was like before X as though it were an explanation of why X happened so late or failed to happen at all. Since the concepts I have mentioned *were* the vocabulary of early scientific thought, it is as misguided to call them an impediment to modern science as to consider walking an impediment to the invention of the automobile.[12]

In the same preface from which this quote has been extracted, Sivin attempts to excuse Needham from the error of having moved in the direction he has just criticized. In a sense, he can do that since Needham has never really sympathized with the internalist theory. The question, however, is whether Needham has not concentrated on the same issue, albeit approaching it from the angle of socio-economic conditions, in other words, using an externalist approach.

To anyone conversant with Needham's work as a whole, there will appear a great deal of truth in the observation made that Needham has often tried to take "anticipations of isolated ideas or techniques of modern science as the sole measure of orientation toward the future". In his restrained critique of Needham's work, Shigeru Nakayama makes the following pertinent remark:

Those who work amid a humanistic tradition tend to think of science as only a minute part of the whole culture. Even today, despite the overwhelming claims of science, even well-educated people seldom have an elementary grasp of its fundamentals. They are not prepared to comprehend the part it plays in their lives. In classical Greece, in Islamic culture, in the Middle Ages, and even in the age of the seventeenth century Scientific Revolution, very few people were interested in and engaged in exact science. Until the establishment of its legitimate position in the nineteenth century, science was only an insignificant constituent in the ocean of general culture; its theoretical basis was an integral part of philosophy. There was no clear-cut scientific culture in the contemporary sense.[13]

If this observation is true and reasonable, it then becomes obvious why Needham has found himself continuously forced to modify his views on the development of science and technology in China. In 1944, in his *Science and Social Change,* he provided his first answer to the question why science of the modern kind did not arise in China. He wrote:

. . . The rise of the merchant class to power, with their slogan of democracy, was the indispensable accompaniment and *sine qua non* of the rise of modern science in the West. But in China the scholar-gentry and their bureaucratic feudal system always effectively prevented the rise to power or seizure of the State by the merchant class, as happened elsewhere.[14]

The real question here is why the social organization of China, having inherited a different range of environmental problems, should and could have been similar to the one that obtained in Europe. In fact, Needham himself has been one of the first scholars to emphasize, in great detail, the originality and distinctiveness of Chinese social and economic organization, law and philosophy, and of course, technology. In *The Grand Titration,* his fifth attempt to resolve the question, he admits that the problem is indeed very complex:

Whatever the individual prepossessions of Western historians of science all are necessitated to admit that from the fifteenth century A.D. onwards a complex of changes occurred; the Renaissance cannot be thought of without the Reformation, the Reformation cannot be thought of without the rise of modern science, and none of them can be thought of without the rise of capitalism, capitalist society and the decline and disappearance of feudalism.[15]

Yet, as Graham has noted, the question remains whether such an analysis still would be able to explain why Galilean science did not arise in China:

We are shown that one of the interlocking factors in sixteenth-century Europe was missing in China, a kind of explanation which is liable to reduce itself to the vacuous observation that conditions in sixteenth-century Europe differed from those of any other place or time.[16]

Graham concludes that we can always think of alternative routes to the scientific revolution, for it is not at all possible to demonstrate conclusively that Galilean science could only have begun in the field of astronomy and that it could not have made its "take-off" with "laws statable in terms of traditional Chinese mathematics, only afterwards refining its geometry to deal with astronomy..."[17] The Dutch historian of science, R. Hooykaas, would agree with an absolutely proper qualification:

The question may be raised whether this result could not have been brought about in a different way. Of course, logically speaking, when *now* a non-christian world manipulates "science in the modern sense", this same situation might have been possible in the seventeenth century with other peoples, and also in other places than western Europe. Historically speaking, however, it makes little sense to reconstruct a course of history different from that which actually took place.[18]

One of the sharpest critics of Needham in recent years has not been Graham, but Mark Elvin, who in his recent work, *The Pattern of the Chinese Past,* presents a different picture of the Chinese merchant community than the one available through Needham's researches. To an extent, this is intelligible, as Elvin has used Japanese sources that escaped Needham. Elvin himself, however, proves unable to escape the great non-question, and finally sets out to provide his own argument about why Chinese science "disintegrated".

He begins by studying the new orientation of Chinese philosophy during the fourteenth century, through the sixteenth, when Wang Yang-ming developed his theory of moral intuitionism, which in effect was a marked shift towards introspection and subjectivity, whereas Chu Hsi, the great synthesizer of Sung dynasty Neo-Confucianism, had urged earlier "seeking for principle in everything". Elvin sees Wang Yang-ming's idealism as a hindrance to the growth of a mechanistic and quantitative approach to phenomena.

Finally, he closely examines the labours of Fang I-chih (1611-71), one of the ablest thinkers in the science of the seventeenth century, and his son, Fang Chung-t'ung, to ask why both of them did not make any real contribution to the development of Chinese science. He is able to conclude that the sophisticated metaphysics of the Fang duo proved inappropriate for good scientific *thinking*: Fang's conviction of the universality of Mind enabled him to solve *any* puzzle in nature through an internal shuffle of Mind:

> Given this attitude, it was unlikely that any anomaly would irritate enough for an old framework of reference to be discarded in favour of a better one. Here then was the reason why China failed to create a modern science of her own accord, and the deepest source of resistance to the assimilation of the spirit of Western science both in the seventeenth century and later.[19]

The trouble with this kind of analysis is that it hardly illuminates; for the point could equally be made that what distinguished Galileo from *his* predecessors was precisely this fact, that they preferred to think in modes that Fang himself would have found quite comfortable. The fact is, however, that Galileo did in reality move further, which means that earlier forms of proto-science do not carry any inherent obstacle to the rise of a person or mind that might think differently.

The researches of R. Hooykaas are relevant here. His boldness in proposing his thesis on the positive role of a religiously oriented culture in the encouragement of the scientific enterprise has now paid off dividends. Recent studies have confirmed the work he published some six years ago.

Metaphorically speaking, wrote Hooykaas, the bodily ingredients of science may have been Greek, but the vitamins and hormones were decidedly biblical. The activities of the Puritans played a crucial contributory role; more important was the influence of Francis Bacon.

It was the Italian scholar, Paulo Rossi, who first drew attention to the millenarian aspect of Bacon's philosophy in his *Francis Bacon: From Magic to Science.* What Rossi did was to show by quotation that Bacon thought of his "Great Instauration" of learning as an attempt to return to the pure state of Adam before the Fall, when, in close contact with God and nature, he had insight into all truth and power over the created world: sin had, since the Fall, clouded his perceptions.

Professor Hooykaas uses the word "Utopian", which is not exactly similar to "millenarian". But he makes a similar point:

> Francis Bacon, though no Puritan himself, had been educated in the spirit of Elizabethan Puritanism, as his religious creed showed, and this spirit was, as Spedding remarked, *incorporated* in his theory of the world. The whole scheme of Christian theology – creation, fall, mediation and redemption – underlay his philosophical works; there was hardly any kind of argument into which it did not at one time or another introduce itself.[20]

Bacon's projected reform of the sciences had thus a strong religious tinge. The primary object of the Great Instauration was "to redeem man from original sin and to reinstate him in his prelapsarian power over created things". The millennium could thus only be brought about through this salvation through science: Bacon himself seems to have believed that the time was short and the End itself not so far off.

In a recent volume, *The Great Instauration: Science, Medicine and Reform, 1626-1660,* Professor Charles Webster makes Puritanism again and its eschatology the ideological framework of the science of the period. He relates the new programmes in education, medicine, and technology of the period to the Puritan effort to restore man's physical perfection lost at the Fall. Writes Frances Yates in a recent review of Webster's volume:

> One would think that people who believed that the End was near would fold their hands and make no further effort. For the Puritans, [however] the millennium had to be worked for with hard social effort, with intense application toward regaining for man the lofty position which he had lost at the Fall. The Puritan doctrine of work was applied to working for the restoration of all things – man and the world would be prepared, through increased knowledge and scientific advance, for a millennial restitution of the state of Adam before the Fall.[21]

This clears up, I think, A.C. Graham's puzzlement, when in discounting the role of time in the development of science, he wrote:

> I must confess to a personal inability to understand why the Hindu is supposed to be paralyzed by the knowledge that no human achievement can outlast a kalpa of 4,000,000,000 years, while the Christian, cramped inside a time scheme of a few thousand years from Creation to Judgement, works hopefully at sciences that have nothing to do with his salvation in the knowledge that the Last Day may already have dawned.[22]

As Frances Yates concludes, the millennium did not indeed arrive, but something else did, and that was the Royal Society, symbol of the arrival of science, "through which man would indeed enlarge his knowledge and his powers, though it has not yet restored him to the Garden of Eden".

With an atmosphere as parochial as this surrounding the early cultivation of universal science, we begin immediately to discover how very misplaced are those learned disputations of historians of science, whether sociological or philosophical, concerning the question why China did not produce similar science. A.C. Graham's "fire argument" may be presented here to clinch the issue. He wrote:

> In the absence of grounds for expectation I explain why a house did catch fire (because someone left a cigarette burning), do not go through all the other houses in turn explaining why they did not catch fire (no one was smoking, the wiring was sound, there were no bombs, no lightning). The difference follows from the fact that like effects may have unlike causes; if the event does happen we can select from the possible causes, if it does not we may not be able to enumerate all the unrealized possibilities.[23]

THE CHINESE PARADIGM

Much of the discussion above would not have appeared in print had historians accepted the obvious fact that as the Chinese had cultivated their civilization in so great an isolation from the rest of the world, the programme dictated by their culture would prove too distinctive to be analysed through Western categories. And what we can proceed now to do is to establish what in effect was the Chinese culture-programme: it was certainly not salvation through science, but it was equally positive and real.

It would be necessary to begin by distinguishing sharply the consciousness of Chinese society from that of its European counterpart. And almost the first reality that we encounter here is that China, first and last, has rarely seen itself under the pervasive influence of a monolithic *religion*: this probably resulted in China escaping the sway of millenarian tendencies.

There were no concepts of salvation or an after-life. In other words, the Chinese have always been oriented towards an immanent, rather than a transcendent, order and it was in the propagation and cultivation of the Great Harmony, an ideal that came through even in the speeches of Mao Tse-tung, that a great deal of intellectual and ethical activity was subordinated. As Professor Kwee explains:

> "Life" is defined as *sing-ming*. The Chinese see life as placed between the poles of the given and the task, potency and realization, origin and destination. Both are considered in the context of the here-and-now, on this earth. Thus, life is not oriented towards a hereafter, as in the neo-Platonian-Augustinian tradition, but rooted in what we understand as Nature.[24]

I think it worth while here to summarize Kwee's essay on *Man and Nature in Chinese Thought,* as an entry into the description of the Chinese paradigm. As Kwee explains, both in the West and in China, man and nature form the two poles of thought. That is, in both Western and Chinese cultural philosophy, and in the pre-modern scientific theories and the social and cultural praxis based on them, one can distinguish a humanistic orientation from a naturalistic one.

However, while in the West, attention was principally diverted to nature, and the natural sciences were developed earlier and further than the human sciences, in China, man formed the focus of both theory and practice. If in the West, man saw himself as able to dominate nature, the Chinese refused that attitude, placing man instead not merely as central, but simultaneously, as an integral part of nature.

Kwee went on to agree with Needham that perhaps the later development of Chinese technology (1500 A.D.) was inhibited when Confucianism, with its strong preference for humanism, prevailed over Taoism, the latter strongly preoccupied with nature. Thus, though both Confucianism and Taoism viewed man as a part of nature, the former devoted its principal concerns to the regulation of human affairs.

> In China, the technology of "natural objects" has not been strong enough to overtake another technology concerning human society itself, a form of

"human engineering" that even in recent times, during the Great Proletarian Cultural Revolution, played an important role.[25]

Here, indeed, we are at the very core of what might constitute a Chinese philosophical anthropology. In China, an individual's *identity* was defined in terms of his harmonious integration in the family structure and in the order of Nature. In other words, the basic focus of awareness was not the individual self-conscious "I" as in the West, but rather the whole structured relationship formed in accord with *li* and the *Tao,* between the souls of the departed and those of the living members of the family.

The basic self-identity is located not in the ego, but in the family ethos. The ideal of harmony means the focus of concentration is not upon the discrete entities themselves, seen in relationship to one another, but rather on the relationship *per se*: the *Tao* is simply the inherent right connectedness of all things. Marcel Granet has even argued that the notion of a soul as a purely spiritual essence is alien to Chinese thought.

This is the Chinese cultural pattern: to criticize it from the point of view of a Western philosophical anthropology is to make some profoundly irrelevant statements. Westerners do not seem to realize that Chinese culture might find no *meaning* in the type of identity demanded by the West for its own members. Criticisms of Chinese totalitarianism are, in my opinion, similarly irrelevant: Since when has the West appropriated to itself the right to determine the proper manner in which another culture might clothe itself?

Kwee concludes his essay by indicating the close connection that exists in Chinese metaphysics between knowledge (of nature) and norm (in society). Nature speaks a language that we can make sense of: this language contains not mere information, but normative structures. Understanding is therefore intimately bound to human transformation.[26]

The object of the Baconian programme was the restitution of society in a millennium that would approximate a condition similar to that before the Fall. The Chinese differed, not so much in the fact that they rejected a millennium: they might conceivably be seen as working towards the restoration of the age of the sage-kings. Rather, and more crucially, they differed about the means or manner in which the Great Harmony might be regained: the normative regulation of human affairs.

In the light of this cultural tendency, the proper representative of the period is not, as Elvin mistakenly took him to be, Fang I-chih, who did make a careful distinction between normative and natural law, only to subordinate the former to the latter; rather is it more contextual to focus on a thinker like Huang Tsung-hsi (1610-95). We are not at all being bold in suggesting that Bacon's *Advancement of Learning* played as crucial a role in the Europe of the seventeenth century as Huang's *Ming-i Tai-fang Lu* (freely translated by de Bary as, *A Plan for the Prince*) was intended to play in the corresponding Chinese period.

The context in which Huang wrote his treatise is not just significant, it is startling. On the surface, this Chinese scholar proposed "in bold terms a new order inspired by traditional ideals".[27] The immediate cause for the work was almost certainly a problem that weighed heavily on his mind, and on the minds of those Ming loyalists despondent after the defeat of the Ming dynasty at the hands of the foreign Manchus: how had this defeat come about, especially when the Chinese had always considered the Manchus numerically and culturally inferior to themselves? Certainly, this was not the first time this had happened in Chinese history. Huang accordingly decided on a thorough re-appraisal of Chinese institutions from ancient times to attempt to discover where precisely the fault lay.

The fourteenth-century writer, Hu Han had noticed that the period of decline had set in after the time of Confucius and had never been halted or reversed ever since. Huang cited this opinion in his own work:

> Since the death of Confucius and throughout the dynasties succeeding the Chou – the Ch'in, Han, Chin, Sui, T'ang, Sung and so on down for two thousand years – the time has not come for a change.[28]

If there was ever the equivalent of the Western doctrine of the Fall in Chinese culture, of a decline from an earlier state of thriving institutions, enunciated by sages, this was probably the one. More interesting is the fact that Hu Han predicted a change after the two thousand years, which coincided with Huang's own period. Huang's work seems to have been inspired by his belief in this announcement of a new Chinese resurgence. *A Plan for the Prince* was to be seen as the blueprint for the re-establishment of earlier harmonies.

There is a parallel to the situation in which Huang found himself in the seventeenth century. During the last reign of the Shang dynasty (1200-1122?), the legendary classical figure, Chi-tzu found himself thrown into prison for criticizing the decadent ways of his king. He was released, however, by king Wu, after the latter had deposed the Shang. Chi-tzu, however, refused to serve Wu, as the latter was really a usurper. Yet, he found himself unable to refuse, when Wu visited him for advice in the running of the country, and communicated to him the political principles that Wu would preserve and practise.

Huang lived in a similar period: except that in his time the Manchus did not come to him for advice (though they tried to patronize him, without success, in later years). There is evidence, on the contrary, in Huang's work, that the new order he envisaged was not to come under Manchu auspices: evidently, he was expecting the rise of a new Chinese power in the near future.

In the *Plan,* Huang first set out a radical examination of the entire range of Chinese institutions:

> Unless we take a long range view and look deep into the heart of the matter, changing everything completely until the ancient order is restored with its land systems, then, even though minor changes are made, there will never be an end to the misery of the common man.[29]

Of course, Huang recognized that many of the proposals made in classical times were inapplicable to the situation in his own day. He is not asking that the past be duplicated, but for a new system of government based on classical principles, which is quite a different thing. Those principles, Huang understood in the light of his Confucian past; they exemplified, one might call it, the Chinese habit of managing affairs through the primacy of the political mean. As de Bary writes:

> For him, as for any true Confucianist, the most fundamental principles are involved in the conception of rulership, since Confucianism asserts that the key to any and all forms of social improvement as well as to all social evils, is the personal example and influence of the king.[30]

Yet the *Plan* remained an eloquent and critical summary of the Confucian political ideal. In it, Huang proposed that the law be made into an instrument more basic than a mere dynastic tool: unlike his predecessors in the Confucian past, he refused to acknowledge that dynastic law was inviolate, though here it should be made clear that the few basic laws he proposed were more in the nature of a constitution or system of government than a proper legal code. De Bary again:

> True law was enacted for the benefit of the people by the sage-kings and is embodied in the system of government laid down in the classics. It consists not in multitudinous statutes, prescribing in detail what men should do and attaching a severe penalty to each infraction, but rather in a very simple and general set of institutions which are basic to proper functioning of government and to promotion of the general welfare.[31]

Huang's institutional reforms included the raising of the status of ministers, restoring the prime ministership (abolished by the Ming), both of which, if accepted, would prove an excellent check to the despotism of the Emperor. He also proposed that the power of the eunuchs be curbed: this could only happen if the Emperor restricted himself in the size of his harem. He also suggested that education and schools be open for all, and provided examples by means of which the civil service examinations might be improved. He asked for a revision of land reforms and land taxation, and for a reform of the lower levels of the bureaucracy.

That Huang's proposals did not see the light of day does not reduce their true originality, so early for his times. The prime condition for the carrying out of those proposals, the rise of a Chinese dynasty, was never fulfilled: the Manchus carried on till 1911, and in the process assimilated what was necessary of Chinese culture for their administration.

Both Bacon and Huang proposed blueprints for a new order; both had the past etched as a model in their minds. Bacon proposed the restoration be brought about through science or knowledge about nature; Huang, in line with his tradition, believed his kind of order would be most effectively created through the active involvement of an enlightened prince. Is it a surprise to observe that the Chinese today still retain the primacy of political action before all else?

Our interpretation of the Chinese and their mind during the period is supported by a number of Sinologists who have been familar with Needham's work about the Chinese past, which they feel contains a number of contradictory attitudes. They note, for example, how Needham, an acknowledged Sinophile, has often praised a great deal of the Chinese past: yet, all too often he has found himself compelled to criticize this very past in the light of his theories of Chinese science and its development. The following is typical of his approach:

In some ways, however, Confucianism was all too humanistic; though humanist, it was anti-scientific. It simply had no interest in the world outside human society. It discouraged such interest.[32]

Nakayama has noted that Needham had already developed his basic scientific viewpoint *before* he devoted himself to Chinese science. More significant is the opinion of Sinologist Arthur Wright, who has remarked that Needham's approach, which makes Taoism the forerunner of modernity, and sees Confucianism and Buddhism as inhibitory factors in the same direction, *misses the point of Chinese culture*.[33] And according to another Sinologist, Arthur Hommel, who is generally friendly to Needham, it was the problems of the world of man and society rather than the natural world that occupied the attention of the Chinese mind, and very little is gained by dismissing them out of hand.[34] This opinion is similar to the one expressed by the Dutch philosopher, Kwee, already quoted above.

All this probably explains why one of the foremost Chinese historians of philosophy, Fung Yu-lan, already in 1922 wrote an article entitled, *Why China has no science*: an interpretation of the history and consequences of Chinese philosophy.[35] Fung understood science as Galilean science: this should be kept in mind when reading the following:

China produced her philosophy at the same time with, or a little before, the height of the Athenian culture. Why did she not produce science at the same time with, or even before, the beginning of modern Europe? This paper is an attempt to answer this question in terms of China herself. . . . I shall venture to draw the conclusions that China has no science, because according to her own standard of value she does not need any.[36]

After summarizing the various tendencies of the different historical schools of Chinese thought, Fung went on to conclude with an examination of the consciousness of Sung Neo-Confucianism:

This period of the history of Chinese philosophy was almost perfectly analogous to that of the development of modern science in European history, in that its production became more and more technical, and had an empirical basis and an applied side. The only, but important, difference was that in Europe the technique developed was for knowing and controlling matter, while in China, that developed was for knowing and controlling the mind.[37]

This extensive discussion of the role of science in culture might turn out for some to be of purely theoretical and academic interest. After the discovery of how to discover, the characteristic of modern science, the value of earlier science seems automatically to be in need of disparagement. Is it? The historian of science, Nathan Sivin, is probably one of few to realize the limitations of that attitude. One of the fertile questions we can ask, he writes in view of our contemporary crisis, "is exactly how science and other aspects of culture co-existed in unity earlier", for they do not do so now.[38]

Sivin has the West in mind in writing that piece, but it should also be obvious that at least two other civilizations have similar problems, from the reverse side so to speak: India and the Arabic lands. No Islamic scholar has expressed the dilemma better than Seyyed Hossein Nasr. Almost the principal thesis of his book on Islamic science argued that no serious study of the Islamic sciences could be carried out without some reference, no matter how brief, to the principles of Islam and the conditions created in time and space by Islam for the cultivation of the sciences, especially when the West slumbered in this department. Not surprisingly, Nasr viewed a possible solution of the conflict raised by the assault of modern science on Islamic culture only in the light of the manner in which the problem had been engaged in past Islamic history. He wrote:

> The Muslims became faced once again in the 13th/19th century with the onslaught of Western science, which has since threatened both the Islamic hierarchy of knowledge and the harmony of its educational system, wreaking havoc with them to an extent that is unprecedented in Islamic history. Al-Fārābi became known as the "Second Teacher" (al-mu'allim al-thānı) because he gave order to the sciences and classified them. To a lesser extent Mir Dāmād performed the same function in Safavid Persia and gained the title of the "Third Teacher". Today, Islam is truly in need of a "Fourth Teacher" to re-establish the hierarchy of knowledge so essential to the Islamic perspective and to classify the sciences once again in such a way as to prevent the sacred from being inundated by the profane and the ultimate goal of all knowledge from being forgotten amidst the glitter of quickly changing forms of science which move ever so rapidly without approaching any closer to the centre of the circle of universal existence.[39]

CHINESE TECHNOLOGY

Needham is quite fond of constantly repeating his claim that, broadly speaking, "Chinese science and technology were very much more advanced than those of Europe between the third century B.C. and the fifteenth century A.D. but that after that, Renaissance Europe began to take the lead".[40] Which makes it all sound like a race that China and the West mutually decided once to contest. Not only that, statements like the one quoted above lead most people to think that after the fifteenth century, technology in China began to stagnate, deteriorate, and decline. Which has in turn, I think, incited Needham, in a recent essay to qualify his earlier pronouncements:

> Perhaps there has hardly ever been so cybernetic and homoiostatic a culture as that of China, but to say that is by no means to speak of "stagnation" as so many Westerners have done; the rate of advance simply continued at its characteristic rate, while after the scientific revolution in Europe change entered an exponential phase.[41]

A better word for the Chinese situation after 1500 is probably "stabilized": it will be evident that by the end of that period, most of the basic technical problems in agriculture and industry had actually *been solved.* It is also inappropriate to term Chinese technology before 1500 as "advanced" as though the Chinese in that period were swayed by the symbols of invention that appeared later in Europe during the Age of Progress. In fact, the Chinese probably saw their principal problem as the weakening of their institutions that had led in turn, as Hu Han realized, to an increasing susceptibility as far as foreign invasions were concerned. This is also the reason, perhaps, why the Chinese treated the mechanical inventions brought in by the Europeans at first as novelties and playthings.

But that the Chinese were mechanically-minded is evidenced by the fact that much before a Polydore Vergil rose in Europe, they had compiled detailed records of inventions and discoveries. The custom, however, rose in Chou times, with the practice of sacrificing to the spirits of the first inventors, who were regarded in fact as "technic deities".[42] Mark Elvin concurs in observing that shrines were often dedicated to the memory of inventors. All these inventors were prized precisely because of the great services they rendered to Chinese society in its early encounter with its environment.

Further, and not unrelated, is the fact that even until the nineteenth century, the Chinese really did not want anything from Europe. Both G.B. Sansom and K.M. Panikkar have made this obvious, but it bears repeating. Sansom goes on to note that not only did China (and India) not feel any great need for foreign merchandise, "they have been under no great inner compulsion to seek wisdom or knowledge outside their own borders".[43]

As Needham himself writes, as late as 1675, the Russian Tsar asked for the services of a group of Chinese bridge engineers; the material influence of China during this period on the West, akin to the influence of Indian textile methods in Europe, but concerning other technical processes has been documented by both Needham and Panikkar and will not delay us here, except to note with Needham, that the Europeans were sending missions of investigators till the middle of the nineteenth century to search out the secrets of traditional Chinese industries, including ceramics, textiles, lacquer, and tea.

Since most of the basic inventions and discoveries had already been in use before the fifteenth century, their description naturally falls outside the scope of this chapter. This in itself is a good thing, since I am then not seriously compelled to cover again the tremendous documentation of Chinese techniques that Needham has achieved in the five published volumes of *Science and Civilization in China*. Even a summary is unnecessary, since Needham has graciously accomplished even that in a small booklet, called *Hand and Brain in China*.[44]

The technological revolution itself in the Sung period has been well treated by Mark Elvin in *The Pattern of the Chinese Past*. It is in fact humanly impossible to keep up with the spate of books and monographs that appear every year. A stray example is the recent monograph by Klaus Flessel, in German: *Der Huang-Ho und die Historische hydrotechnik in China* which, incidentally confirms the points I have already made regarding Chinese technology before the fifteenth century.[45]

Flessel shows that the technical aspects of hydraulic engineering involving the river had been emphasized, studied, and solved before the end of the T'ang dynasty, and that the hydraulic works served their purposes so well that "there were no further fundamental innovations in hydraulic structures until the introduction of Western technology and the use of concrete as a building material". What did remain were administrative problems, which were ultimately solved during the reign of the Sung dynasty, and these involved the centralized coordination of these hydraulic works.

Yet the talk of "stagnation" goes on. Arnold Pacey, in his essay on the role of ideas and idealism in the technology of China, writes that after the middle of the fifteenth century, a tendency was slowly gaining ground in China to feel that the practical arts needed little further development. He tries to explain this by emphasizing that Chinese intellectual life was becoming less sympathetic to some forms of technical development and that in some fields, the practical arts were already entering a period of stagnation.[46]

I think it is time to throw this notion of China "stagnating" right out of the windows of our historical minds, and there is no better way of aiding the process than by presenting a summary review of one of the largest industries in China during the period of the Ch'ing: cotton. What we shall say about cotton may equally be said about the Chinese silk industry during the same period. We can anticipate our conclusion by observing that both these industries hardly showed any great signs of "stagnation".

In his comprehensive essay on *Cotton Culture and Manufacture in Early Ch'ing China*, Craig Dietrich suggests, after a summary outline of the history of the cotton industry in the Ch'ing period that there was in China,

> a capacity for change that is at odds with the stereotype of timeless China. By the time of the Ch'ing, change had slowed, and no revolutionary mechanical or organizational development appeared during that period. Indeed part of the industry could be described as simple and changeless. But another part reveals itself to be differentiated and adaptive. The industry possessed, for example, a range of techniques and organizational forms that permitted cloth to be made both by self-sufficient families and by a system of market-oriented specialists.[47]

Though cotton was known in China since early times, it had never been considered more than a kind of exotic commodity. And it actually reached its position as the single most important fibre in the Chinese economy only during the two or three centuries that corresponded with the late Sung, the Yuan (the Mongols), and the early Ming.

The reason for its new popularity seems to be, as happened in England a little later, the pressure of population on land resources. Though we are to deal with this issue at leisure below, it seems proper to mention already at this stage that by 1050 A.D. the pressure of population had already resulted in the de-forestation of large portions of North China. This in itself may not seem so extraordinary a revelation, till we discover that plant yields for cotton per acre are ten times those of other fibres, including hemp, which cotton soon replaced, and which together with silk, till that time, constituted the basic material for clothing. The introduction of cotton in such a context would be immediately favoured, for it would supply two important functions: release land for food production, and at the same time meet the needs of a larger population and even trade. Mark Elvin concludes that the cotton gin's arrival in the thirteenth century in China was a reasonable response to the sudden glut of raw cotton, and quotes Wang Chen's *Treatise of Agriculture* that appeared in 1313:

> In times past rollers were employed (to remove the seeds from raw cotton). Nowadays, the gin is used. . . .It is several times more advantageous than the use of rollers. Even if there is a large quantity of cotton, the use of this method permits one to get rid of the seeds immediately, and to avoid a backlog piling up.[48]

An interesting fact here is the activity of the Taoist nun Huang, who is said to have brought more efficient techniques for ginning and spinning cotton from the Hainan Island into the lower Yangtze area; so grateful were the people of the village of Wu-ni Ching, near Shanghai, where she settled, that in the typical Chinese fashion, they erected a shrine to her.

By 1760, the use of cotton had spread all over China; the perennial variety of cotton tree grew in the more suitable wet climate of the remote Southern areas, while the annual shrub variety took deep root in the heartland of China itself and in the Yangtze region: the latter forming the basis of the newly developed Chinese cotton industry. As Li Pa, prefect in Fukien province, wrote that year:

> If we search for (the fibre) that is most widely used, that is most reasonably priced and labour-saving, that is suited both to cheap and to expensive textiles, that benefits rich and poor alike, (we will find that) only cotton has all these exceptional qualities. . . . In all the places that my feet have left their traces there was no man who did not wear cotton and no soil that was not suited to its production.[49]

Dietrich concludes on the basis of his own studies that between three-fifths and four-fifths of all *hsien* (country), both in the late Ming and in the early Ch'ing periods, manufactured some cotton cloth. In contrast, however, to the period of dynamic growth and change earlier when cotton was introduced on a wide scale, the industry seems to have stabilized, even with regard to its geographical distribution and specialization. The technology developed had apparently reached a plateau: the limits to any possible improvement in technique had already been reached by the end of the Ming period.

What China was experiencing in this period was the introduction and switch to a new resource base in its textile industry: the crucial point is that China adapted to it, not only with new techniques of production, but novel methods of social organization. Something now about both these aspects of the industry.

As early as the Sung, the Chinese had discovered the utility of a small iron roller with which to accomplish the process of ginning: the removal of seeds from cotton fibre. The double-roller soon after made its appearance, from India. Early in the fourteenth century, two cranks were used to move the rollers. Still later, a treadle was attached to one of the cranks. Some time before the eighteenth century, a fly-wheel was added to the treadle-powered roller to sustain its rotation.

For loosening the cotton fibres, since they normally got matted through transport or package, the Chinese exploited a device, also used earlier in India (and still used in the latter country today) similar in construction to an archer's bow, and with a hard, tough, and taut string. When struck, the string vibrated and reduced the matted cotton to a fluffy state, free even of impurities. Later, these bows got larger and had to be vibrated through the rhythmical strokes of a mallet.

The spinning of this fleecy fibre was accomplished on a series of different devices. The simplest of these, in Ch'ing China, was the suspended spindle, still unexcelled for producing fine and even yarns. But it was not long before the spinning wheel arrived, originally developed in India again in the thirteenth century, to supplant the suspended spindle which lacked its ease and speed. The single-spindle wheel remained the norm all over Ch'ing China, except in the densely populated lower Yangtze area, where the greater demand for yarn soon stimulated the invention of the compound wheel, which exploited three spindles. Actually, the Chinese experimented with two, four, and five-spindle apparatuses, but finally settled for the three-spindle machine as it proved to be the best compromise between quality and quantity.

As with the spinning operations, so with the weaving processes: a new range of instruments was soon made available, depending of course on the relative size of the operation. Different kinds of spools, reels, frames and cranking devices were used to combine "several yarns into long, multiple-ply threads suitable for warp". And to prevent warp threads from the friction of the weaving process, various methods of sizing were evolved for the purpose.

By the time of the Ch'ing, there were three main kinds of looms: the horizontal, the "waist" and the draw. While the first was in general use, the last was generally reserved to produce luxury textiles, which involved the weaving of elaborate patterns and designs.

The Chinese dyeing and printing processes are described in Stuart Robinson's two volumes on the subject and will not delay us here. The arts of dyeing and printing were jealously guarded secrets, evidently of great economic value in a thriving industry, and they were only passed down from father to son or trusted apprentices. The Chinese craftsman was also proficient in the art of resist-dyeing and printing with wood-blocks. And when it came to giving the final product a glossy finish, the art of calendering was not behind the others in any sense. Crang Dietrich concludes:

> Thus in two or three centuries the Chinese had adapted and improved the gin, the bow, and the spinning wheel, all imported from South Asia, and combined them with appropriate Chinese devices such as reels, looms, and calendering stones. The result was a technology suitable for a range of productive organizations, from the single family to highly differentiated, market-oriented entrepreneurs.[50]

An adapting technology made possible an equally flourishing trade, which in turn forced the industry to be increasingly commercialized. "Thus, even if there is a bad harvest in our counties, our people are not in distress so long as the other counties have a crop of cotton."[51]

Those peasants that were drawn within the commercial network normally fulfilled varying functions. Some merely grew cotton and then sold it all to the brokers. In other areas, as in Wu-hsi *hsien,* the people worked on cotton brought from elsewhere. This might be partly explained by the fact that in some areas at least cotton refused to grow; a good example is the frustrated attempts of the villagers of Jui-chin *hsien* in the Kiangsi area. Significant in this context is the beginning of cotton manufactures in north China in the seventeenth century, so far impeded by the tendency of the threads to snap in the dry atmosphere: now the problem had been overcome by spinning the yarn in moist underground cellars and the local supplies that were normally sent to the Yangtze region for processing, were now kept back for the industry, so that the people of Yangtze had to look for fresh sourcess of supply from Manchuria.

The commercialization of the cotton industry led to increasing specialization and differentiation of functions. In some areas, brokers who bought large amounts of cotton, found it profitable to begin an independent ginning industry, since the removal of seeds lessened the weight of the cotton bags and made for cheaper transport costs. In other areas, ginning and bowing processes were combined under one roof, and the end product then sold to spinners. There were still others who began to specialize in the bowing industry itself.

Spinning for commercial purposes could also only be done in specialized units. One late Ming gazeteer suggests a strong economic reason for this, when he notes "that the poor people lack funds and cannot weave cloth. Daily they sell several ounces of yarn to make a living." As for specialization in weaving processes, this could only be possible if a certain amount of yarn was produced specially for the market. "A late-nineteenth-century source for a *hsien* in Kiangsiu notes that the northern and eastern villages had established themselves as producers of high-quality yarn." This was sold to villagers in other areas, who wove it into good quality cloth, again for the market.

The finishing processes, of their very nature, demanded a specialized social organization: they required materials like dyes and sophisticated equipment. As for calendering itself, the Su-chou area in the lower Yangtze region employed many thousands of workers to process large quantities of cloth. What follows is a description by Li Wei, Governor-General of Chekiang, in 1730 – it provides a picture of tremendous activity, a thriving market, great inter-regional flows of goods and a complex social organization:

> In the prefectural capital of Su-chou. . .the green and blue cotton cloth from the various provinces is bought and sold. After it has been dyed it has to be given lustre by being calendered with large stone foot-rollers. There is a class of persons called "contractors" who make ready large stones shaped like water chestnuts, wooden rollers, tools and rooms. They gather together calenderers to live there, and advance them firewood, rice, silver and copper cash. They receive cloth from the merchant houses to be calendered. The charge per length is 0.0113 of an ounce of silver, all of which goes to the aforesaid workers. Each of them, however, gives to the contractors each month 0.36 of an ounce of silver as representing the rent for the workspace and the tools. . . . Formerly there were only seven or eight thousand men in the various workshops. . . . Now a careful investigation of the area outside the Ch'ang Gate of Su-chou has shown that there are altogether over 340 persons acting as contractors, and that they have set up more than 450 calendering establishments, in each of which several tens of thousands of men are employed. There are over 10,900 calendering stones, and the number of workers must equal this.[52]

As I said earlier, a similar case could be made out concerning the active texture of the silk textile industry in the period of the Ch'ing. An excellent summary description is available in E-Tu Zen Sun's comprehensive article on *Sericulture and Silk Textile Production in Ch'ing China,* to which the reader is referred.[53] Right now, we can conclude with Mark Elvin that

> the last three centuries of pre-modern Chinese history saw the creation of much larger units of private economic organization than ever before, and that the change here was qualitative as well as quantitative. In particular, rural industries were co-ordinated through a network of rapidly increasing density, and urban industry, supplied with materials and customers through this network, developed new structures to handle larger numbers of employees.[54]

The fact is similar descriptions may be drawn not merely concerning cotton and silk, but of other commodity production system in China. Mark Elvin provides two excellent pre-modern descriptions by contemporary writers of the large activity in the porcelain and iron industries. Most of the material has been brought together in a recent bibliography of science and technology in China, to which the reader is referred.[55]

Arnold Pacey, with many others including Needham, has come to feel that "the classical Confucian motif of scholarly austerity" as practised by Chinese administrators and scholar-landlords was perhaps responsible for a withdrawal of Chinese interest not merely from, for example, the marine adventures of the period round 1400, but from other areas of practical concern as well.

I have tried to suggest that there are two ways in which we might approach the central role of the Confucianists in the life of Chinese society. We might discuss them from the point of view of modern science, in which case we would feel it necessary to express dismay about them and to befriend the Taoists. I have disagreed with this approach, observing that Chinese society should be seen on its own terms. The second approach is a natural conclusion of seeing the first as wrong, and suggests that merely the refusal of the Confucianists to entertain the preoccupations of the Taoists does not imply that they had begun to grow increasingly introspective.

On the contrary, against the Taoists, the Confucianists re-emphasized the primary element of their long tradition: the concern and regulation of human affairs. It is therefore in this light that we proposed the crucial importance of a work like Huang Tsung-hsi's *Plan for the Prince.* For, if there is anything for which Huang is known today, it does not concern his being an original thinker or the founder of a new school, both of which he was not; it is because he was able to combine the broad scholarship of the Chu Hsi school *with the active interest in contemporary affairs* that characterized the best of the Wang Yang-ming school.

In our current age, enthused as most of us are over the phenomenon of a scientific culture, it is indeed difficult for us to imagine that some future historian of human affairs might assume a criterion of judgement that would give precedence to those societies that concerned themselves more with men and action (Hanna Arendt) than to those that attained a state of unusual dominance through the exercise of a method that, by postulate, ignored the human. As Fitzgerald expressed it in *The Birth of Communist China*, such an historian may suggest

> that technical skills are not good criteria of true civilization, that harmony and balance in a human society are better than restless change and the chemical search for progress to some undefined goal.[56]

CHINESE POPULATION AND THE TECHNOLOGICAL SYSTEM

In the final part of this chapter, I shall examine the role of the fourth element of our *homo faber* model: the factors outside Chinese societal and imperial control that gradually and inexorably questioned the stability that that society had succeeded in preserving right up to the early decades of the nineteenth century. I am not very much interested in the question why China did not produce an industrial revolution: in fact, when we have finished, it might seem all the more obvious that the question does not make any sense.

My analysis here concerns the socio-economic causes that led to the fall of the Ch'ing dynasty in 1911. No historian worth his salt, however, would and should accept the overriding value of such "impersonal" interpretations of events. Yet that they are often influential over the course of actual events is not very easily deniable. History seems to be a continuous shift between the poles of a society being controlled and its ability to control or cope.

What Chinese society was not able to control was the growth of its population. I shall examine the close relation between population growth and the pressures it exerts in the direction of agricultural and technological development in the chapter that follows. F · now, I merely assume the hypothesis.

In a recent article, co-authored with Ray Huang, Needham wrote that "it is easy to over-stress the influence of philosophers and to minimize the effect of concrete environmental and economic factors. . . . The importance of ideology, however great, can never obscure the basic fact that underneath lie the material forces of climate, geography and social integration," a point I made in the first chapter.[57]

Needham and Huang note that the high degree of centralization that developed very early in China was not an invention of political thinkers, but rather inspired by circumstances, geography being one of the crucial factors. (One of the most startling sketches of geographical influence on the nature of Chinese society is the map made by Chi Ch'ao-Ting, concerning the principal economic areas of China, and attached to the first volume of Needham's monumental work).

In their discussion on the influence of the river, the Huang-Ho on Chinese life, both writers stress the fact that the emergence of a unified China in 221 B.C. was favoured by the need for hydraulic works, not only in flood protection, but also for irrigation, and later, for bulk transport. Thus, though it should not be overemphasized, the river and the Chinese need to control it played an important part in the political organization of China for many centuries.

Chinese population pressures and the response of Chinese society to it might be understood in much the same terms: again, what surely emerges is the ingenious Chinese responsibility to cope with this gradually increasing weight of numbers. But there is always a point of diminishing returns.

The classic work on the population of China, based on Chinese records, and yet to be updated, is the book of Ho Ping-Ti, *Studies on the Population of China, 1368-1953*.[58] Some revisions of Ping-Ti's figures are available in Perkins' *Agricultural Development in China: 1368-1968*,[59] which specifically investigates the manner in which Chinese agriculture evolved to meet the increasing needs of a growing population. Elvin's recent work has also a discussion on the same subject.[60] I shall briefly present the skeletal framework of this entire issue, first in general, then in the specific case of the cotton industry: this will facilitate our resumption of it in the following chapter on the industrial revolution in England.

The first signs of a disequilibrium in the relation of a population to its environment are, as Braudel summarized them, the occupation of new territory, emigration, the clearing of new land, agricultural improvements, and urbanization.[61] I have already indicated the deforestation of China in the north about 1050 A.D., the clearing of new land. But already in the fourth and fifth centuries, Chinese settlers had begun to migrate in large numbers towards the south, in the direction of the Yangtze river valley.

I follow Esther Boserup here in her classic analysis of the stimulation of agricultural development.[62] The methods of cultivation, at first, in the Yangtze region were crude: the land was cleared by fire, then flooded, and eventually abandoned.[63] The increase of the population, however, reduced this earlier system of long-term fallow cultivation, a characteristic of most sparsely populated areas. The population became more settled, and wetfield rice cultivation, as opposed to the dry farming of the north, soon turned out to be the most favoured method of cultivation. As Perkins writes:

> Accompanying the expansion of rice cultivation were the development of new tools, new crop rotations, and hundreds of new seeds. Wet rice, transplanted into flooded fields, required a radically different technology from that used to grow millet and wheat on the parched land of the north.[64]

Thus, we can agree with Elvin's observation that Chinese agriculture saw itself transformed between the eighth and the twelfth centuries. It is useful, at this stage, to point to the difference in the trends of agricultural development in Europe and China.

Population pressure gradually shifted the economic centre of gravity away from the Mediterranean in Europe: the movement was towards the north, and the symbols of advance were the axe, the heavy plough, and the efficient horse harness. Lynn White has also shown how the introduction of the new, heavy plough to work the heavier soils of the north necessitated a change in the very shape of the fields.[65] The Chinese advance, as noted, was towards the south, to a river valley with few forests, but with problems of equal significance, in this case, concerning irrigation. The corresponding symbols of advance included the dam, the sluice-gate, the noria, and the treadle water-pump. Both advances could only be made through the expenditure of a great deal of labour, a situation accepted only under conditions of great population pressure.[66]

By the time of the Northern Sung (960-1126), the Chinese population was already over a hundred million. The population of England at the beginning of the industrial revolution was about 6 million (1740). The difference in the scale of magnitudes would normally wreck any attempt at meaningful comparisons: yet these comparisons have repeatedly been made.

Mark Elvin probably comes close to the truth when he sets out to describe the changes this large population brought about as a series of revolutions in farming, water transport, money and credit, market structure, urbanization, science and technology. There is not much point in my presenting even a brief view of the medieval economic revolution, as he calls it: more relevant for us at this stage are the conclusions of Perkins to which we can now turn.

Perkins' studies of the six centuries of pre-modern and modern China allowed him to conclude that even though China's population increased five- or six-fold during the period, Chinese farmers were able to raise grain output, and that they did so in more or less equal measure through expanding acreage and raising yield per acre.

The Mongol invasions had greatly reduced the earlier increase in population, but by 1400 the population had begun to rise again. With a temporary lapse at the fall of the Ming and during the Taiping rebellion in the middle of the nineteenth century, it continued gradually to increase from around 80 million in 1400 to about 583 million in 1953. Yet the expansion in grain output continued to more or less match the increase in numbers. How was this possible?

First through migration into uncultivated land: even till after the fourteenth century, the southernmost regions of Kwangtung, Kweichow, and Yunnan were able to receive large numbers of migrants. In the early part of the Ming, this migration continued towards areas like Kwangtung, but also now into the sparsely populated areas of central China: Hunan and Hupei. Later, migrants found a further opportunity to move onto the North China Plain and the northwest: these trends continued through the eighteenth century. In the Ch'ing period itself, the major recipients of migrants were the provinces to the west, particularly the southwest. In the twentieth century there was still some land left, poorer land no doubt, but that too soon filled up. Most of this land was in the northeast, the home land of the Manchus and in some regions in the northwest.

But Chinese farmers were also able to increase grain yields on existing land, through a variety of means: technologically, however, the elements of rice production had been developed and spread to the populated areas long before the fourteenth century. The improvements came with other means, including the planting of a large number of new seeds, the organization of the system of manure, in which hogs played a great role. As Perkins notes, "the rise in the number of hogs and draft animals made it possible to more or less double the application of fertilizer per *mou* between the fourteenth and twentieth centuries". All in all, through sheer ingenuity in increasing the efficiency of existing means, changing cropping patterns (which was possible through extra labour), and the occupation of new land, the Chinese agricultural system was able to prevent a decline in food intake: per capita grain consumption remained stable through the six centuries.

The moment there was no more land to open up, diminishing returns began to set in, brought about by factors related to the land. With pre-modern techniques, Chinese wheat yields even in the 1920s were substantially above those available in most of Europe on the eve of the industrial revolution: fourteen bushels of wheat per acre to the 9.5 bushels in France at the end of the eighteenth century. Rice yields for the Chinese in the 1920s were in the region of 56 bushels.

Investigations in the 1930s in the regions of Shantung and Hupei suggest that the land in these places could have profited from larger quantities of manure. The shortage of manure was, however, due to a shortage of grazing land, which in turn reflected the need of a dense population to turn pasture into arable.

Thus a civilization that in the third century B.C. and even up to the Sung had used livestock as an important element in the farm economy, was by the late traditional period and the twentieth century forced to adopt vegetarian ways. Even the hogs were incapable of improving matters: their increase depended on an increase of grain. W.F. Wertheim, after his visit to China in the fifties, could write:

> The civilization of China is essentially based on vegetables. Clothes, utensils, houses, all of them are made principally of vegetable materials. About 98 per cent of the calories in the Chinese diet are of vegetable origin.[67]

This picture, of a steady movement of the Chinese population towards a kind of ecological wall, has a corresponding equivalent in Chinese traditional industry. Few people are willing to accept the fact that in such a vastly populated country, the traditional textile industries of cotton and silk, and the encouragement of traditional means of production in them, were not merely a necessity, but a positive boon, enabling as they did the average Chinese agrarian household to achieve a better means of subsistence. Mechanization in such a context could only have been seen as a crime: Elvin provides instances of "Ludditism" in China in this context. The issue is better pursued through an analysis of the Chinese cotton industry's fortunes.

I have already noted that cotton replaced hemp in the thirteenth century, and the sudden glut of cotton that arrived, stimulated the invention and use of the cotton gin; later, the ready supply of labour would disfavour further invention on a radically wide scale. The Jesuits in 1777 had already seen the problem:

> The question of the utility of machines and working animlas is not so easy to decide, at least for a country where the land is barely sufficient to feed its inhabitants. What use would machines and working animals be there? – to turn part of the inhabitants into philosophers, that is to say into men doing absolutely nothing for society and making it bear the burden of their needs, their well-being, and what is even worse their comical and ridiculous ideas. When our country folk (the argument is expounded by Chinese Jesuits) find themselves either supernumerary or unemployed in a few cantons, they decide to go away and work in great Tartary, in the newly conquered countries where our agriculture is making progress. . . .[68]

By the seventeenth and eighteenth centuries, most of the poor in China (as in England with the woollen industry) were to be found employed in the spinning and weaving of raw cotton. They made daily trips to the market to get this raw material. We have already noticed that a great deal of this industry was based on subsidiary labour: income from spinning and weaving constituted only a portion of the total income of a peasant household. Thus, for a part of the year, when agriculture demanded all attention, even the simple equipment lay unused. If demand rose, then the huge capacity in hundreds of thousands of peasant households was brought into play by diverting labour from agriculture. If demand fell, the damage was not so serious, as it affected only a portion of the total composite income of each peasant household; even this might be alleviated by re-directing labour into farming. Thus, in times of boom there were no great prospective rewards for inventors, even no need for them. In times of slump there were few penalties severe enough to drive the inefficient permanently out of business.

Even such a situation, beneficial in the absence of anything better, could last as long as population did not continue to rise; yet it did. By the sixteenth century, for example, there was already

Even such a situation, beneficial in the absence of anything better, could last as long as sixteenth century, for example, there was already an acute shortage of cotton in the lower Yangtze valley, the most densely populated area in China and also the main centre of the cotton industry. Inter-regional flows of cotton began on a massive scale, signalling a breakdown of local self-sufficiency. Land that had been reserved for cotton was now needed to grow grain; and by the sixteenth and seventeenth centuries, there was little land available in China for any crop except foodgrains.

In fact, any expansion in the supply of raw cotton beyond bare parity with population growth depended on raising higher the per acre agricultural productivity that was already the highest in the world, or, alternatively, on increased imports of either cotton or food. Thus, between 1785 and 1833, the *single* province of Kwang-tung imported on average from India, each year, six times as much raw cotton as all Britiain used annually at the time of Arkwright's first water-frame invention.

Now, with the wisdom of hindsight we might look back as it were and wonder if mechanization of the cotton industry might have helped. Mark Elvin unfortunately does this, when he observes that the persistent difficulty of obtaining raw materials (unlike England's case) cannot have made the creation of *labour-saving* machinery seem an urgent necessity, as though the English industrial revolution was itself a matter of labour-saving machines – it being decidedly not. Elvin adds that the Chinese in fact possessed the knowledge to mechanise textile spinning if they so desired.

He finally sums up this concatenation of events as a "high-level equilibrium trap". In his words:

> With falling surplus in agriculture, and so falling per capita demand, with cheapening labour but increasingly expensive resources and capital, with farming and transport technology so good that no simple improvement could be made, rational strategy for peasant and merchant alike tended in the direction not so much of labour-saving machinery as on economizing of resources and fixed capital. When temporary shortages arose, mercantile versatility, based on cheap transport, was a surer and faster remedy than the contrivance of machines. This situation may be described as a "high-level equilibrium trap".[69]

That all this is a great deal of sophisticated nonsense will be seen in the chapter that follows.

CHAPTER FOUR

English Technology and Culture: 1500-1830

> All of you who are sixty years of age can recollect that bread and meat, and not wretched potatoes, were the food of the labouring people; you can recollect that every industrious labouring man brewed his own beer and drank it by his own fireside. You can recollect when every sober and industrious labourer that was a married man had his Sunday coat, and took his wife and children to church, all in decent apparel; you can recollect when the young men did not shirk about on a Sunday in ragged smock frocks with unshaven faces, and a shirt not washed for a month, and with their toes pointing out from their shoes, and when a young man was pointed at if he had not, on a Sunday, a decent coat upon his back, a good hat on his head, a clean shirt, with silk handkerchief round his neck, leather breeches without a spot, whole worsted stockings tied under the knee with a red garter, a pair of handsome Sunday shoes which it was deemed almost a disgrace not to have fastened on his feet with silver buckles. I appeal to you all, those of you who are sixty years of age, whether this be not a true description of the state of the labourers of England when they were boys.
>
> –William Cobbett: 1820

If man is everywhere disposed to technology how is it possible that in the Britain of the period under discussion an attitude to technology appeared that was out of keeping with earlier attitudes and which was responsible for the erection of the mammoth structure of the industrial revolution?[1] The absence of such an attitude seems to be conspicuous not merely within the civilizations of China, India, and Islam, but even within the nations of Europe. Europe, contrary to all that has been written on the subject, did not produce the industrial revolution, Britain alone did: any attempt to found the revolution on European culture as such must flounder on the fact of this obvious divergence.

In fact, it was only a hundred years after those phenomenal beginnings on the island, that the countries on the Continent found themselves capable of catching up with the lone industrial pioneer (aided, of course, by the fact of their relative backwardness and the ready availability of a model they had merely to imitate).

The answer to the question will itself throw light on the appearance of a new scale of values that came to dominate the civilization of England, and which exhibited itself, to be precise, in the preference of English man for consumption over leisure. After the revolution, the demands of culture came gradually to be fulfilled through an extension of the productive system over non-economic areas of life. In this sense, British society proved to be not merely the workshop of the world, but also a laboratory in which far-reaching results first appeared.

It is this edge that consumption achieved over leisure that has led me to continue using the terms "the industrial revolution" to characterize the processes through which the age passed, even though I am quite aware that a number of historians have cautioned against their appropriateness. There has been no doubt an unwarranted assumption of the emergence of industrial, economic, and technological processes during the period as sudden and somehow spontaneous. T.S. Ashton, who concentrated on the economic aspects of the industrial revolution, emphasized their phenomenality with a great deal of caution and qualification.[2]

Of the other historians who have spent some time on the issue, R.J. Forbes comes closest to the views expressed here. In *The Conquest of Nature*: Forbes wrote:

> The Industrial Revolution... was by no means as sudden as is often claimed, nor as revolutionary as some have believed. It had its roots in the important technological changes of the 16th century, although it did not gain momentum until about 1800. From a social point of view, however, the changes during the period from 1730 to 1880, dramatic in their strange medley of good and evil, often tragic in their combination of material progress and social suffering, might indeed be described as revolutionary.[3]

The chronological conventions concerning the industrial revolution first proposed by the elder Arnold Toynbee in his lectures on the subject, and published posthumously in 1884, have been further eroded by the weighty evidence brought up by the sixteenth-century English historian, John Nef, who holds that the British economy had already assumed industrial proportions in the period 1570-1640, when production found itself oriented to the needs of the general population, and not merely to those of a select few, as earlier:

The growth in the importance of mining and manufacturing in the national economy was, it seems, scarcely less rapid between the middle of the sixteenth century and the Civil War than between the middle of the eighteenth century and the first Reform Act.[4]

Such a sense of continuity from earlier periods is missing, for example, in the recent, now standard, rather academic treatment of the industrial revolution, David Landes' *Unbound Prometheus* I prefer to see, with W.F. Wertheim, a fundamental difficulty with those interpretations that imply that a cataclysmic change can occur within the bounds of a stable and equilibrium dominated society.[5]

ON THE NATURE OF TECHNOLOGICAL REVOLUTIONS

As we begin with Britain in 1500, we are confronted first with an agricultural civilization, supplemented no doubt with a large manufacturing capacity, and in nature, quite transitional. So much has been written about what happened *after* the transition: little about what occurred before. Every work on the industrial revolution carries a compulsory brief account of an agricultural revolution that preceded, or ran simultaneously with it. A more thorough scrutiny of the agricultural scene at this early date will lead to the conclusion that the word "revolution", when seen from the angle of technological development, exaggerates the extent of the changes involved.

It is necessary at this stage to entertain a detour and to sketch out the skeletal features of a framework within which a stricter description of English agriculture would be possible. In dwelling for a while on the nature of what constitutes an "agricultural revolution", we might be able to exorcise a few notable, even influential, fallacies. The ensuing discussion will also aid, I feel, a better understanding of non-industrial attitudes to work, to efficiency, and to the proliferation of goods; it might also explain the despair of development experts, driven by the Western experience of agricultural work and economic activity, when faced with peasants, farmers, and tribals in the Southern countries of the world.

One strong and influential view of change, held by such notables as Carlo Cipolla and Lynn White, sees progressive changes in agriculture as the result of autonomous technical revolutions: the resulting surplus created is then supposed to lead in turn to an increase in population.[6] Those who follow this line of thought have normally set forth to discover numerous types of agricultural "revolutions", each enabling a greater surplus and a greater population.

Historically, and even in recent times, few events have reinforced such a view. Few observers, for example, would like to suggest that the tremendous growth in population rates witnessed throughout the Southern countries in the two post-war decades could be explained as a result of changes in the conditions of food production. As I have tried to show in China's case, the rule normally is for a population to be pressured by its increasing numbers to assume the exploitation of more intensive or newer techniques for the increase of output. The Danish economist, Esther Boserup, made a detailed study of this issue, and it is her hypothesis that I shall now re-propose.[7]

One of the major contentions of Boserup is that the growth of population is a major determinant of technological change in agriculture, so that one can normally expect some correlation to exist between the rates of population growth and that of technological change. Conversely, one should be able to observe technological regress after the reduction of population.

Boserup notes that primitive forms of agriculture use land unintensively and yet provide the population with adequate subsistence and with little hard work. The earliest form of agriculture, slash-and-burn, involves little more than clearing an area in a forest through fire and ring-barking the trees to kill them. Seeds are then cast into the ashes or digging sticks used if tubers are to be planted. Burning destroys the seeds of potential weeds. All fertilizing is unnecessary due to the system of fallow used: the land, after use, is let alone to regain its fertility, usually for a period ranging from twenty to twenty-five years, before being used again. Such a system is only possible if there is plenty of forest land and the population quite sparse.

Indeed, one of the great surprises of recent anthropological research is a falsification of the long-held view that primitive economies have always been on the brink of starvation. The anthropologist, Marshall Sahlins, even went so far as to label Palaeolithic hunters (before they were even required to adopt slash-and-burn) as the original affluent society.[8] Further, anthropologists have come to recognize that early societies, contemporary primitive cultures, and pre-industrial economies not yet compelled to radical change under population pressure, are often characterized by what one calls a "leisure preference". In other words, once a society of this kind and number has taken care of its subsistence needs, it turns its attention to its non-economic activities, or rather refuses to allow the former precedence over the latter. Joan Robinson discusses a number of such economies in one of her recent books.[9]

To return to Boserup, as population density increases in a particular area, and more people lay claim to land, the extensive use of land, including the permissibility of long fallow periods, has to be necessarily curtailed. In the case of slash-and-burn agriculture, the peasant can make do with an ordinary digging stick; in the following phase, he is compelled to invent and use a hoe. *The hoe is not an advance in technics: it is a new tool for a new technical situation.* Since fallow time is shortened, the forest no longer has sufficient time to regain its original state. Bushes now become the norm.

Now, bushes do not prevent sunlight, as do trees, from reaching the grasses at floor level: weeds grow. Weeding is a nuisance arriving only in this new circle of events: in some cases, the labour involved in weeding might equal the labour required for clearing new forest land. The total amount of labour in bush-fallow is therefore doubled. Moreover, under bush and shrub vegetation the soil itself tends to become more compact: original forest soil under the cover of dead leaves does not. Neither the digging stick nor the plough can here do the work exclusively restricted to the hoe. Further, productivity per unit of land *falls* under bush-fallow; this is probably the reason why primitives refuse to exploit bush-fallow if the easier system of forest-fallow is still available. The Peru Incas had evidently been in this state of cultivation in the mountainous Andes: to criticize them for not having invented the plough is therefore wrong, for the plough was not required in their circumstances.

For ploughing becomes a necessity only after population density has further reduced fallow time till only grasses can grow. Grassland cannot be prepared for cultivation by burning: the turf will survive the fire, it must therefore be turned. *The plough is again no advance on the earlier system of hoe and bush-fallow:* it involves on the contrary more labour and greater technical problems. Draught animals become a necessity and any draught animal requires a larger area for grazing than it can plough. The ingenuity of *homo faber* is evident in the balance he has been able to create between the amount of land set aside for grazing, the number of animals needed for ploughing and eating, and the supply of manure produced for the fertility of the arable land.

But further population growth may even break this equilibrium.

Notice the situation in large parts of India and also North Africa, today, where population increase in this century has contributed to a sharp shortage of pasture and a corresponding worsening of the condition of animals. The European agricultural revolution in the middle ages faced and solved this very problem: it was a response in fact to a grazing shortage. The "revolution" consisted in changing from a system of short fallow cultivation to one of annual cropping, with fodder plants as part of the new system.

Some of the plants were leguminous crops that fix nitrogen directly into the soil from the atmosphere, increase its fertility. The revolutionary aspect comprised the rise in output, plus fodder plus a surplus. Yet few historians of technics have taken count of the labour costs the new system demanded, which were nearly double those of the earlier system. Fodder crops are support crops, but require as much labour as the annual crops like wheat. In other words, the farm labourer had now to work twice as much as he had done before. Whereas earlier he had had to work merely some months in the year, he now had to labour all the year round.

It is precisely this large increase in labour hours that farmers refused to accept as long as population pressures did not require it: all the general features of the revolution had in fact been known long before. The methods introduced in this period, including the significant innovations of crop rotation without fallow and the use of leguminous plants as fodder were used in the ancient world in the Mediterranean as well as in other densely populated regions of the world. Boserup writes:

> Their reappearance began in the densely peopled and highly urbanized valley of the Po, and from there they moved to England and Northern France via densely peopled and urbanized Flanders where the turnip, for instance, was used already in the thirteenth century. These few facts suggest that the transition in Europe from short fallow to annual cropping was not the result of contemporary inventions: it could more plausibly be described as the spread of various methods of intensive cultivation most of which, although known since antiquity, were little used in Europe until the increase of urban population raised the demand for food and the increase in rural population provided the additional labour needed for a more intensive cultivation of the land in the most densely populated regions of the continent.[10]

POPULATION AND ENGLISH AGRICULTURE

If we return to England, we observe three broad population

movements in its history of the past ten centuries. The first rise occurs during the middle ages and continues till the mid-fourteenth century. The rise is countered by Malthusian checks and the Black Death. The population begins to rise again after the middle of the fifteenth century till about 1640, when it stabilizes at roughly five and a half million. The third rise begins about 1740. As M.M. Postan has shown, these figures are shaky: it is best to supplement them with other non-statistical indices, as land-reclamation.[11]

But in each period the relation between technological development and population increase is obvious. That the increase comes *before* renewed technological change is evident from the study of wages: wages tend to fall during periods of population increase and rise in periods of population decrease or stability. The causal role of population growth behind the sixteenth-century price inflation has been proved, for example, beyond doubt by F.J. Fischer of the London School of Economics.[12] Also, this period of growth fits in with Nef s description of an increase in the volume of manufactures and their scale, specially if we keep in mind the fact that before the industrial revolution manufactures provided supplementary incomes for surplus agricultural labour.

A comparison with China in the same period is instructive. The Chinese population provided for increased technological change during the eleventh and twelfth centuries, stimulating even the construction of polders. The number of people was radically decreased during the Mongol invasions. It began to rise again in 1400 and continued at a progressive rate, with a few setbacks caused by the fall of the Ming, epidemics, and the Taiping Rebellion. The areas of greatest technological change, more intensive systems, including multiple cropping, were also those of the densest populations. But notice, for example, the near absence of technological proficiency in the same period in the Hupei province, observed by visitors from the more highly populated areas:

> For the most part, the fields in Hupei are different from those in Chiang, Che and Min (the east and southeast) . . . . The land is so sparsely peopled that they do not have to bestow any great effort on their farming. They sow without planning or weeding out. If perchance they have weeded they do not apply manure, so seeds and other sprouts grow up together. They cultivate vast areas, but have poor harvests.[13]

In England, the population growth of the twelfth and thirteenth centuries brought about the colonization of new lands: wooded areas were cleared and marshes and fens drained and reclaimed. Once this colonization was no longer possible, the farmer turned to a more intensive utilization of existing fields. At this stage the law of diminishing returns began to set in. The average fertility of land and the yields from seed began to decline during the thirteenth century; pasture could no longer be increased to provide greater services to arable land. Fallow periods were considerably shortened, poorer land used, increased yields attempted by liming, and ploughing in straw ash. But these improvements did not, as in China, keep pace with population increase.

This is evident in at least two periods, 1160-79 and 1300-19, when wheat prices more than tripled. The result approximated more closely to the situation that exists in the Southern countries today: poverty. The real need for supplementary sources of income stimulated manufacturing activity and produced, what one economist in a flight of enthusiasm, has termed, "an industrial revolution in the thirteenth century". Most of this activity was centred round the staple industry of England: wool, which was expanded, so that by 1310 English exports of wool had increased by about 1,040 per cent. The shift of the wool industry from the older urban areas to the countryside confirms the increase of rural involvement, and this in turn favoured the development of fulling machines which required running streams.

The decline of population after the Black Death led to a corresponding regress in agricultural activity, and to a lack of rural labour. Throughout the land, the progress in material wealth which had been so marked a feature of the reign of Edward I (1272-1307) had not only been arrested; civilization and refinement were lost and England at the accession of Henry VII (1485-1509) was far behind the England of the thirteenth century.

By the middle of the sixteenth century, however, population had begun to increase again and to provide the conditions of agricultural growth. One of the effects was a rise in the price of food and in the value of land, which meant greater wealth for the landowners and those able to buy land, and absolute destitution for others. In 1538 already, a German chronicler could write that "landed property and rents for dwellings have become so very dear that they can hardly go higher". And a generation later, an Englishman repeated the complaint:

The people are increased and ground for ploughs doth want, corn and all other victual is scant. People would labour if they knew whereon; the husbandman would be glad to have ground to set his plough to work – if he knew where.[14]

These uprooted landless labourers soon aggravated the condition of the labouring classes in the urban areas: surplus labour had its effect on wages. By 1571, in the building trades, for example, wages could buy only two-thirds as much as the wages of a century before, and this fall in the standard of living would carry on for another forty years.

No issue reveals the state of affairs more ideally than the problem of enclosures: the increased value of land led to its being fenced, principally by powerful landowners. Enclosures had begun as early as the thirteenth century, chiefly due to the profitable trade in wool, which necessitated protected pastures. The descriptions concerning the issue make this obvious; Thomas More's *Utopia* contains references:

sheep. . . these placid creatures, which used to require so little food, have now apparently developed a raging appetite, and turned into man-eaters . . . nobleman and gentleman, yea, and certain Abbottes . . . leave no ground for tillage: they enclose all in pastures; they throw down houses; they pluck down towns; and leave nothing standing but only the church, to make of it a sheep house.[15]

A mid-century Protestant preacher complained that "whole towns are become desolate, and like upto a wilderness, no man dwelling there, except it be the shepherd and his dog". And an anonymous writer set out to demonstrate the decay of England through an argument which maintained that "every time a plough was put out of use by the increase of sheep, six persons lost their employment and seven and a half persons their potential supply of bread". A John Hales wrote in 1549:

. . . where XL persons had their lyvings, now one man and his shepherd hath all. . . . Yes, those shepe is the cause of all these meschieves, for they have driven husbandrie out of the countries, by the which was encreased before all kynds of victuall, and now altogether shepe, shepe.[16]

As Robert Heilbroner notes, the process would not stop until the mid-nineteenth century. He notes that in 1820 the Duchess of Sutherland dispossessed 15,000 tenants from 794,000 acres of land, replaced them with 131,000 sheep, and by way of compensation rented her evicted families an average of two acres of submarginal land each.

"Depopulating enclosure" was exposed, condemned, and combated by a long series of proclamations, royal commissions, and statute laws, yet it carried on because few dared interfere with the liberties of the powerful and dominant landowning class. Poverty now came to be seen as a social problem; a large population on the move could only be identified as a potential menace. For wage-earners who had no other source of income – a class which must have included a high proportion of all those who left the land at this time, hoping to better their prospect in the towns – the evidence all goes to show that it was a more difficult period than almost any other in English history. Unlicensed beggars and vagabonds, seen by the upper classes as a direct threat to society, were believed to be about 10,000 in number in early Elizabethan times.

Enclosures increased during the industrial revolution, though the motive behind them this time was no longer wool, but food. About 17 years after the third spurt in population, that is, in 1757, England, a food exporting country, now became the reverse. Demand exceeded supply and could only be fed through imports. The period also saw the *quadrupling* of the price of wheat. As prices arose, the profits of landowners moved in the same direction. It had become profitable in other words to improve agriculture and output.

Writes Ashton:

> Some writers who have dwelt at length on the fate of those who were forced to leave the land have tended to overlook the constructive activities that were being carried on inside the fences. The essential fact about enclosure is that it brought about an increase in the productivity of the soil. Many of those who were divorced from the soil were free to devote themselves to other activities: it was precisely because enclosure released (or drove) men from the land that it is to be counted among the processes that led to the industrial revolution, with the higher standards of consumption this brought with it.[17]

This is evading the issue, which is that technological change to improve the productivity of the soil came about at the expense of the lower dispossessed groups; that the motivation to produce more was not to feed the supplanted but profit, in a situation where food had become a profitable commodity. And the result was that the displaced were not "free to devote themselves to other activities", but compelled to wander in villages and towns asking for alms, or to enter the cities to become the unwilling pawns in a factory system where labour was demanded for lengths of time longer than had ever been the case in agriculture.

This is not to deny that in some enclosures, landowners, using more intensive techniques, required more labour than before, as Landes has pointed out.[18] The fact is the land had been in the first place misappropriated. Besides, not all those who had enclosed land actually set out to improve it, and it is this fact that eventually turned the mind of one of the most ardent supporters of the enclosure system: Arthur Young. A tour he made of East Anglia in 1800 shocked him, and convinced him that the entire system had been rigged by the landowners for their own benefit, which if cultivated would have produced enough food to keep the poor from starvation or from the workhouse. But so little did the landowners care that often the poor did not even know who the land belonged to: certainly it did not any more belong to them.[19]

The writing of agricultural history has never been achieved by those who were actually required to make it possible: by the large majority of the poor before and during the industrial revolution. Lynn White once wrote that the urban roots of the word "civilization" are more evident in the neglect historians have lavished on the rustic and his works and days. "Not only historians but documents in general were produced by social groups which took the peasant and his labours largely for granted."[20] Yet White himself follows his complaint with the following:

> To be sure, we have heard that in the late seventeenth and eighteenth centuries "Turnip" Townshend and a few other adventurous agronomists in Britain and on the continent developed root and fodder crops, reformed agriculture, *and thus provided the surplus food which permitted labour to leave the fields* and to man the factories of the so-called industrial revolution (emphasis added).[21]

Finally, it should come as no great surprise to note that one of the most influential treatises of the late eighteenth century concerned itself with population. In 1798, Thomas Malthus published *An Essay on the Principle of Population as It Affects the Future Improvement of Society,* in which he set out his famous law concerning population and resources:

The power of population is infinitely greater than the power of the earth to produce subsistence for man. Population, when unchecked, increases in a geometrical ratio. Subsistence increases only in an arithmetical ratio. A slight acquaintance with numbers will show the immensity of the first power in comparison with the second. By that law of our nature which makes food necessary for the life of man, the effects of these two unequal powers must be kept equal. This implies a strong and constantly operating check on population from the difficulty of subsistence.[22]

Before the time of Malthus, in medieval England, in fact, a kind of unwritten pact had existed between Church, landlords and serfs, which had been designed to mitigate the hardships arising from what was then the commonest form of involuntary poverty, the removal by death or disablement of the family breadwinner.

The Black Death broke the stability of this system, but in 1601, Elizabeth I passed an Act in which the landowners and other well-off classes were legally committed to pay rates for the maintenance of the poor. The switch from alms-giving to rate-paying soon led to a change of attitude on the part of donors: if by their contribution they were no longer laying up for themselves treasures in heaven, there was nothing to be gained by generosity. Inevitably, they began to look at the poor through the eyes not of Christians, but of taxpayers.

When Malthus arrived, this rate-paying or poor relief had come to a stage where donors regarded it as nothing more than a millstone round the neck of Progress. Evidently, poor relief could not be justified on economic grounds. And if it should be demonstrated that a large population was *not* in the national interest, the final justification for the poor law would be removed. And in 1798, Malthus set about removing it, in his Essay.

He claimed that ordinarily a check operated on any population growth. As food shortages would result, the price of food would rise; but as the supply of labour would be increased, wages would fall. Inevitably, many of the poor would suffer severe distress and their numbers would be soon reduced. The poor law merely perpetuated the problem: it was a futile exercise calculated to perpetuate misery. The poor, Malthus argued, had no *right* to relief; a man born into an already full world,

if he cannot get subsistence from his parents, on whom he has a just demand, and if society does not want his labour, he has no claim of *right* to the smallest portion of food and, in fact, has no business to be where he is. At nature's mighty feast, there is no vacant cover for him. She tells him to be gone, and will quickly execute her own orders, if he does not work upon the compassion of some of her guests.[23]

Malthus, and the neo-Malthusians today, would believe that the supply of food for the human race is inherently inelastic and that this inelasticity is the main factor governing the rate of population growth. Thus, scholars like Cipolla, White, or Gordon Childe [24] see population growth as a dependent variable, determined by preceding changes in agricultural productivity which, in turn, are to be explained as a result of extraneous factors, such as fortuitous technical inventions. Those who see the relationship between agricultural change and population increase in this essentially Malthusian perspective agree that there is at any given time in any given community a warranted rate of population growth with which the actual growth of population tends to conform. Unfortunately, there is no evidence at all for such a view. Whatever evidence we have proves the contrary, that more mouths mean more hands and more production.[25]

An economist looking for the roots of the situation that leads to increased economic growth would find them in the breakdown of self-sufficiency, the increase of trade, of manufactures, and innovations in agricultural technology. He will then proceed to universalize this experience until he realizes that there have always been societies that, being in a state of ecological equilibrium, their populations controlled through social and cultural restraints (as Malthus realized), are not forced to start out on the road to economic development in the first place. To put this idea into the jargon of economics, innovation, economies of scale, education, capital accumulation and so on are not causes of growth; they *are* growth. The causes of growth would include at least up to and including the industrial revolution, the dynamic role of population pressures.[26]

As Richard Wilkinson observes, if there are some societies that do not wish to sacrifice existing levels of efficiency for increased production, others may be equally unwilling to sacrifice possible increases in leisure. It is possible that some societies will thus use improvements in techniques to reduce the time needed to produce subsistence, in order to increase their leisure. Economists (and a large number of "development" economists), who have been trained to attach a higher priority to increasing the output of goods, would have preferred the extra time being exploited for increased production.[27]

THE INDUSTRIAL REVOLUTION

Which brings us to the industrial revolution and the role of population growth in its appearance. A perusal of the literature might bring the reader, not aware of the larger context, to the conclusion that here at least *homo faber* entered a new phase of security in which machine power took over not merely functions reserved for man, but multiplied results many times over, to enable a standard of living rarely before achieved in human history. The "official" dates demarcating the period of the revolution are 1760 and 1830. However, the most important innovations concerned resource changes, and these came much before 1760; and no industrial revolution would have been possible without them.

Let me first begin by parading my thesis. I am convinced that the process of technical change is not inherently developmental; every technical change that is necessitated by the circumstances of the period raises a host of new problems, most of which are unceremoniously dumped on the backs of the weakest members of the particular social group. Histories of inventions, taken by themselves, mean little for they tend to present the unwholesome, unrealistic view that technology is one of the few branches of civilization in which the line of development is always on the upgrade, that mechanization itself undergoes constant evolution, and that technical change always follows the principle of efficiency and moves always forward. "While faiths veer and fashions oscillate, techniques advance", is all too common a wisdom. In such a context, efficiency becomes the sole criterion of progress and those preoccupied with it will proceed, conceptually and methodologically, to detach the process of technical change from its historical, geographical, and human context.

It is not that I do not see any merit, contrary to some scholars, in the appearance of the industrial revolution. On the contrary, here is evidence that man was once more able to encounter and solve problems on a scale and using methods for which there is very little historical precedent. But this itself indicated that the industrial revolution was a response to a wide range of problems which we discover as having appeared in a society gradually facing a state of ecological disequilibrium and with no traditional solutions.

In adapting to these problems, English society was coerced into making corresponding fundamental changes in its pattern of life. Perhaps, this should be qualified: it was the majority of the English people, unsupported by privilege, those that constituted the lower rungs of English society, that were forced to accept, for the sake of subsistence, a cultural deprivation and a disruption of symbols that till today has not seen repair. The minority rich were rarely called upon to face this, so I shall not concern myself with them.

Richard Wilkinson summed up the context of the industrial revolution in ecological terms: the initial stimulus to change came directly from resource shortages and other factors connected with these, and their effects on an economic system expanding to meet the needs of a population growing *within a limited area*. The emphasis is important if we are to distinguish the *English* roots of the industrial revolution. The continental countries were also undergoing increases in population, but were hardly restricted by limited areas to find solutions. In England, as traditional resources reached scarcity levels, it became necessary to substitute new ones in their place, new technology to process them, and longer working hours:

> The ecological background to the industrial revolution was an acute land shortage. In the centuries before industrialization the English population was dependent on the land for almost all its materials. The supply of food and drink depended on agricultural land, clothing came from the wool of sheep on English pasture, and large areas of land were needed for extensive forests: almost all domestic and industrial fuel was firewood, and timber was one of the most important construction materials for houses, ships, mills, farm implements etc. In addition, the transport system depended on horses and thus required large areas of land to be devoted to grazing and the production of feed. Even lighting used tallow candles which depended ultimately on the land supply. Land was bound to become in increasingly short supply as population increased.[28]

I have already discussed the land question, so we can move on to tackle the fuel and energy problem posed by the most important land-based resource after land itself: wood. People living in gas-supported societies tend to forget that wood constituted the source of everyday power before the industrial revolution. Note Braudel, who writes:

> Calculations of power today leave out work by animals and to some extent manual work by men; and often they ignore wood and its derivatives as well. But wood, the first material to be in general use, was an important source of power before the eighteenth century. Civilizations before the eighteenth century were civilizations of wood and charcoal, as those of the nineteenth century civilizations of coal.[29]

The shortage of wood in England was first felt sharply during the population increase of 1540-1640. The fact is easily illustrated if we point out that it was during this period that the price of firewood rose nearly three times as fast as the general run of prices. A part of the shortage was due to the increased production of iron (charcoal) and the needs of the shipbuilding industry; a larger part was due, however, to the increasing number of people that needed wood for domestic purposes. The "timber famine" was therefore rooted in ecological disequilibrium: the conversion of woodland into arable, and the necessity for larger quantities of domestic fuel. The import of wood did not help matters. In 1631, Edmund Howes wrote down the experience of his times:

> Within man's memory, it was held impossible to have any want of wood in England. . . . But. . .such hath been the great expense of timber for navigation, with infinite increase of building houses, with great expense of wood to make household furniture, casks, and other vessels not to be numbered, and of carts, wagons and coaches, besides the extreme waste of wood in making iron, burning of bricks and tiles, that at this present, through the great consuming of wood as aforesaid, and the neglect of planting of woods, there is a so great a scarcity of wood throughout the whole kingdom that not only the City of London, all haven-towns and in very many parts within the land, the inhabitants in general are constrained to make their fires of sea-coal or pit-coal, even in the chambers of honourable personages, and through necessity which is the mother of all arts, they have late years devised the making of iron, the making of all sorts of glass and burning of bricks with sea-coal or pit-coal.[30]

"*Coal* grew almost entirely with the number of urban – and especially metropolitan – fire-places," writes Eric Hobsbawm. "Since the quantities of coal burned in British homes were very much greater than their needs for iron. . . the pre-industrial base of the coal industry was much sounder than that of the iron industry."[31]

Thus, it was during the pre-industrial period actually, due to the paramount timber shortage, that many industries that needed large amounts of fuel in heating processes, including glass-making, salt-boiling, brewing, and brickmaking, switched to coal, a mineral resource. The shortage of wood, however, would affect the production of a large number of tools, instruments, and containers made usually from it, leading thus to a greater demand for iron.

The making of iron itself, however, depended on wood. Iron and glass making factories often had to shift their locations from forest to forest. A blast furnace built in Wales in 1717 was not fired until four years later, when "enough charcoal had been accumulated for thirty-six and a half weeks' work". The blast furnaces that had replaced the "bloomery" technique associated with small forges could, because of the perennial lack of wood supplies, operate only once in two or three years, or even sometimes a year in five, seven, or ten. Calculations show that an average iron works where the furnace worked two years on and two years off needed the resources of 2,000 hectares of forest before the eighteenth century. It is not surprising to find increasing legislation during this period to preserve forests: the navy too needed timber.

The serious shortage that developed could only be offset by imports of iron from Sweden, where forests were in plenty and the cost of the smelted metal quite low. And this dependence on Swedish imports would continue for a round' sixty years after the Restoration: the English production of iron remained static and low. And it was this concern that finally drove men to attempt to harness coal to iron production. One man, Dud Dudley, made the point so obvious:

> I have held it my duty to endeavour . . . the making of iron . . . with pit coal, seal coal, peat and turf, for the preservation of wood and timber so much exhausted by iron works of late.[32]

It was only in 1709 that Abraham Darby thought of borrowing the solution to the maltsters' and brewers' problems and used coke for iron-smelting, but a generation was to pass before this method was refined through knowledge and experience to make it more acceptable for general use, and still another generation before coke-blast iron could be directly converted into wrought-iron.[33]

The dilemma faced by historians who see the substitution of wood by coal as a conscious choice of sorts exercised by British ingenuity in its all-encompassing striving for progress is to explain why coal was not used earlier on a larger scale, when it was already known; unless the answer is that most people saw coal as an inferior fuel compared to wood, as, in the food department, they considered the potato as inferior to bread loaves.[34] Further, where wood was plentiful, it continued to be used. In areas not so well favoured, the change to coal was inevitable.

In industries, for example, where substances were kept separate from the fuel in vats or containers, it was quite easy to substitute coal for wood. But in processes like metal-smelting, where the fuel came into contact with the raw material, or in drying processes where things were hung in the fumes above the fire, coal could bring about undesirable chemical changes in the product. Precisely as the heavy plough had once demanded a new shape to fields, coal forced bakers, for example, to change the design of their ovens to avoid contaminating their bread with coal fumes, and brick-makers had to experiment long and hard till they found the less gaseous coals that did not fuse the bricks together, glass-makers had to use covered pots, and maltsters had to develop further the use of coke to avoid the smoky gases and tars given off by raw coal.

Coal is also disadvantageous as a domestic fuel due to the harmful constituents it contains; and its widespread use for domestic purposes during the late sixteenth and early seventeenth centuries was paralleled by the spread of chimneys as the smoke forced people again to abandon the traditional custom of having fires in the centre of the room, with a hole in the roof.[35] The rich were able to delay coal use much longer than the poor, just as they could continue to wear wool while the poor had already been reduced to cotton.

It is more appropriate, therefore, to see the use of wood instead of coal in the preceding centuries as a rational choice in a situation where wood was plentiful, and therefore cheap.[36] The switch came about when timber scarcity raised the price of wood till it was more expensive than coal; in most countries today in the Southern hemisphere wood is still more inexpensive than coal. Further, the use of coal involves high transport cost. In places where open-cast mining is no longer possible, deep mines raise costs again as production encounters new technical problems. Such costs in England at the time of its resource change became hidden costs: only a traveller from a wood-economy entering what had turned into a coal-economy would have noticed the difference.

Seen any other way, the issue will generate contradictions in any standard history of technology. David Landes provides a good example. At one place he is required to admit that coal, unlike wood, was not the best of all possible fuels. He writes:

> From the sixteenth century on, as we have noted, the need for new sources of thermal energy in a country almost denuded of its forests led Britons to substitute mineral for vegetable fuel in a wide variety of heat absorbing industrial operations. At the same time, the consumption of coal for domestic purposes rose steadily: there was perhaps a time, in the sixteenth century,

when the Englishman recoiled at the acrid, sulphurous fumes of burning coal; but by the modern period, such scruples were laid aside by familiarity and necessity.[37]

Landes' commitment to progress in industry for its own sake is patent in his discussion of the comparative advantages that prevented coal usage on an English scale on the Continent:

On the supply side, the contrast between Britain and the Continent was less sharp. Yet the resources of the mainland countries were in fact less favourable to industrial expansion than those of Britain even before the change in raw material requirements consequent to the Industrial Revolution. The cloth industries of France, the Low Countries, and Germany, for example, had to import the bulk of their fine wool from abroad. And the lack of concentrated, easily accessible known deposits of coal led to a neglect of the possibilities of mineral fuel; here, indeed even nature's bounty hurt, for the *relative* abundance of timber seems to have encouraged retention of the traditional technique.[38]

So we should not be surprised that Arthur Young was amazed to discover that "the wheels of these (French) waggons are all shod with wood instead of iron". I have also mentioned the significant fact that coal did not overtake wood in the American economy until the 1880s, even though the use of coal by then had long been established in industrial processes. Landes goes on to produce a bundle of contradictions, in writing:

Whatever the sources of this ferruginous temper, it is the more impressive for having developed in the face of the growing scarcity of fuel; until well into the eighteenth century, Britain used iron because she wanted to, not because it was abundant or cheap. (To be sure, the most likely substitute, wood, was perhaps even dearer.) Even so, one can but wonder what would have happened, had she had to go on depending on costly and inelastic foreign sources for much, if not most, of the principal structural material of modern technology.[39]

It is when we consider the history of the steam engine that we can fully understand the strength of the ecological argument. It is well known that the steam engine was the invention of Hero of Alexandria: the library at Alexandria contained a perfectly working model of it. Greek society might have ignored it for a variety of reasons. The West, however, slept over the design for well-nigh two thousand years. When it was resuscitated, it was not for the purpose of exploiting sources of power on a large scale as it is expressed in textbooks, but to raise water from mines that had reached below the water table.

The problem arose as the demand for coal increased, and after open-cast deposits had been exhausted: mines had to be sunk deeper and deeper. Mines on the hillsides might be drained by digging a special drainage shaft that led out of the hill to a point below the level the mine had reached. But once below the water table, the problem of drainage became acute. Mines close to streams could exploit a water-powered pump; otherwise, a horse-whim might serve the purpose. But the use of these methods was restricted to shallow depths. By the end of the seventeenth century, however, depths up to 200 feet were common in most mines, and some had even reached 400 feet. At such depths, horse-whims, bucket pumps, and rag-and-chain pumps began to suffer the law of diminishing returns: more energy was spent in moving the machinery itself, little left over for lifting the water.[40]

Attempts to raise water through some means of "fire" had been tried out since 1631. Steam power seemed to be a worthwhile proposition not because it was initially more powerful than horse or stream power, but because the power itself could be delivered in a more appropriate form: one had either to reduce the pressure above the water to be raised or increase the pressure below it, and, in either case, coal to fuel the steam engine was available in plenty, at the pithead itself. Horses displaced meant that less fodder needed to be grown in a period of land shortage. Further, not all mines could be located, or found themselves located near streams.

Thomas Savery's "Miner's Friend", patented in 1698, unfortunately blew up too often. Newcomen's "fire-engine" was available only in 1712. Even so, only one was in operation in England thirty years later, in 1742. In the following thirty years, a further sixty were in use in the tin mines of Cornwall. For nearly a century, thus, the use of the steam-engine remained confined to a reciprocating pumping notion. And only in the late eighteenth century, when the new cotton mills began to demand rotary power, did Boulton and Watt succeed in manufacturing the first steam engine harnessed to produce a rotary motion. The reason for this delay was perhaps due to the continued existence of a number of sites where water-wheels provided the easiest and most economical means of obtaining a rotary motion for mill production. The increasing acceptance of the steam version later must also be put down to the fact that not only was it a dependable source of power, but also because unlike a stream, it could be moved from place to place. Till that time, however, the rotary steam engine did not displace traditional methods in the performance of traditional tasks.

Richard Wilkinson has surveyed the evidence to confirm his hypothesis that a similar situation did lead to innovations and resource substitutions in other important, basic areas of the English economy. The development of transport during the period was stimulated, for example, by the breakdown of self-sufficiency in two vital areas: deficient communities now had to import fuel and grain, sometimes metal, and people who had formerly produced commodities themselves were now forced to enter into increasingly complex trading relationships. Secondly, the shortage of agricultural land made horse transport increasingly prohibitive, for pasture entailed the sacrifice of arable land, which was already in short supply. The growth of wheeled transport led to better roads, an invention in itself not required earlier, and the rise of turn-pike trusts to finance their development.

The mania for canal construction arose directly in response to the horse problem: the high costs of hay and corn increased the costs of goods. This is more easily accepted if we know that the feeding of each horse required the amount of hay grown on between four and eight acres of land. Traders were aware of these problems at the time. As Wilkinson notes:

> An engineer writing in about 1800 on the proposed Grand Survey Canal Navigation calculated that "as one horse on an average consumes the products of four acres of land, and there are 1,350,000 in this island that pay the horse-tax, of course there must be 5,400,000 acres of land occupied in providing provender for them. How desirable any improvement that will lessen the keep of horses. . . ." The Earl of Hardwick writing in favour of the Cambridge and London Junction Canal used a similar argument: "If the canal should be the means of releasing 1000 horses from. . . employment. . . 8000 acres of land . . . might be applied to more useful purposes, which would help to keep the labouring poor from suffering from want of bread.[41]"

Arguments to counter the high initial costs of the steam railway were expressed in similar terms. Witness the following report to the House of Commons on "steam carriages" in 1833 and its calculations:

> It has been said that in Great Britain there are above a million horses engaged in various ways in the transport of passengers and goods, and that

> to support each horse requires as much land as would upon average support eight men. If this quantity of animal power were displaced by steam-engines, and the means of transport drawn from the bowels of the earth, instead of being raised upon its surface, then, supposing the above calculation correct, as much land would become available for the support of human beings as would suffice for an additional population of eight millions. . . . The land which now supports horses for transport on turnpike roads would then support men, or produce corn for food and the horses return to agricultural pursuits.[42]

Timber shortages had their effects on the material used in house construction, if we keep in mind the fact that at least till coal replaced wood in brick kilns, it required more wood to build a house in brick than to build it in timber. Coal firing enabled extensive brick production without any increase in unit costs. After the Great Fire of London in 1666, legislation to secure rebuilding of houses in brick was made correspondingly easier.

Resource changes had important consequences in the early chemical industry. Alkalis were important for a wide variety of products including glass, soap, alum, and saltpetre. When the different processes used wood, potash was readily available as a by-product. The switch to coal made potash supplies scarce; they had to be imported. This led to the processing of large quantities of sea-weed. The chemistry historian, N.L. Clow, has summarized the industry's earlier development "as a subsidiary facet of the search for an alternative to wood".

Ecological pressures also lay at the back of the transition from tallow candles to gas lighting. Tallow candles were obviously dependent on land supply. Prices rose in the industry sufficiently high to warrant the import of large quantities of raw material. In 1838, Britain still imported over £1 million worth of vegetable and whale oil to satisfy home needs. Yet, gas lighting was known already at the end of the eighteenth century. Why did the change now come about earlier? Simply because people were being asked to use an inferior product: the fumes from the impure gas were so unpleasant that most people preferred to go on using the older oils. When gas use became inevitable due to high costs of the oils, users went so far as to mount the lights outside the windows so as to avoid the fumes.

It is the textile industry, however, that carries a few important lessons in the actual development of the industrial revolution and which links the latter with the technological experience of colonial areas. Central to the issue is the reason why cotton, not wool, was the first to undergo mechanization, specially if we know that wool constituted the principal staple industry of the English economy up to the eighteenth century. Historians who set out to describe the "conscious and progressive" nature of the industrial revolution seldom get beyond this dilemma. David Landes has tried to explain it away by pointing out that wool-mechanization was more difficult than that of cotton, but his arguments probe rather than convince.[43]

In the period before the industrial revolution the woollen industry formed the foundation of English manufactures. As one description has it:

> Wheresoever any man doth travel, ye shall find at the hall door. . . the wife, their children, and their servants at the turn spinning or at their cards carding, by which commodities the common people live. . . . The weaver buyeth the yarn of the spinster, the clothier sendeth his cloths to the tucker or fuller, and then the merchant or clothier doth dye them in colours, or send them to London or elsewhere to his best advantage.[44]

Technical innovations had been tried out in the industry before: the stocking frame was delayed by the hostility of the hand-knitters and so was the "Dutch loom". And in 1733, when John Kay invented the "fly-shuttle", it was opposed, according to some writers, because of the "conservatism" of the workers themselves.

As in China and India, the methods and scale of the woollen industry had reached a stage where a precarious balance existed between social, ecological, and economic constraints. The spinners and weavers were often from among the poorest in the community and took up their trade when their land holdings proved insufficient. One study has in fact shown that the woollen industry tended to concentrate in areas where inheritance customs led to the successive division of land in periods of population growth: villages that practised partible inheritance tended to be more densely populated than others; land holdings were smaller and people had to earn part of their incomes from weaving and other such work.

The price and quantity of wool supplies (unlike those of cotton) were largely fixed by the land situation; markets were limited, and given the technology, comparatively competitive. In such a context, the inefficient machinery in use (inefficient, of course, in the light of further possible refinements) was maintained as a way of ensuring a sufficiently wide distribution of the small rewards available. Both the early inventions that transformed the cotton industry, Kay's fly-shuttle and Wyatt and Paul's spinning frame, were invented with wool in mind, not cotton. In the context, hostility to them in the woollen trade was therefore to be expected.

Legal action often supported workers in the trade, but it would soon prove helpless in preventing radical changes brought about by ecological disequilibrium. More people on the land meant simply less pasture for sheep, especially when this was complicated further by the new profitability of food itself. As prices rose and real wages fell, a new resource become necessary for clothing materials for the majority of the population. Cotton would have been used early, as it was known, if people had but desired it, but in a cold climate it was considered inferior to wool. Scarcity made the difference.

As opposed to the seasonal demand called up by fashion, cotton first came to be used by the poorest customers for general purposes all the year round, and being cheaper its widespread use would soon be favoured in the context of high wool prices. In the middle of the nineteenth century, Engels made the point obvious when he described the clothing of the working class:

> Linen and wool have practically disappeared from the wardrobes of both men and women, and have been replaced by cotton. Men's shirts are made of bleached or coloured cotton cloth. Women generally wear printed cottons: woollen petticoats are seldom seen on the washline. Men's trousers are generally made either of fustian or some other heavy cotton cloth. Overcoats and jackets are made from the same material—Gentlemen, on the other hand, wear suits made from woollen cloth. The working classes . . . very seldom wear woollen clothing of any kind. Their heavy cotton clothes, though thicker, stiffer and heavier than woollen cloth, do not keep out the cold and wet to anything like the same extent as woollens.[45]

More significant in the case of the cotton industry, however, is the fact that its supplies did not have to be grown on British soil and thus compromise an already aggravated land shortage. Raw cotton could be imported in large quantities first from the Indies, and later, after Eli Whitney, from America. During the American Civil War, India was made to step in as substitute supplier.

Arnold Pacey's figures for cotton imports into England speak for themselves. From 1760, that is, twenty years after the third population rise, to 1775, cotton imports increased from two to seven million pounds. Within the next fifteen years, they increased from seven to thirty-five million pounds.[46] As with China, with the gin, in England too, the crucial inventions appeared much after the demand for cotton had raised imports to large proportions. Hargreaves' Jenny appeared in 1765, Arkwright's water frame in 1769, Crompton's mule in 1779. In no case did the invention precede the increase in demand.

I have claimed that periods of transition and resource changes brought about by population pressure on limited land are more often than not periods of intense discomfort to the majority of the population; not only have the lower classes to make do with inferior products, they are also compelled to work more and keep longer hours. This fact was no doubt noticed by some contemporaries of the period.

The historian, Rowland Prothero, after a study of documented enquires, speeches, and pamphlets on the subject of the agricultural labourers' distress during the period 1800 and 1834 (he was writing fifty years later), concluded that their standard of living had sunk

> to the lowest possible scale; in the south and west wages paid by employers fell to 3s.–4s. a week, augmented by parochial relief from the pockets of those who had no need of labour; and insufficient food left its mark on the physical degeneracy of the peasantry. Herded together in cottages which, by their imperfect arrangements, violated every sanitary law, generated all kinds of disease, and rendered modesty an unimaginable thing . . . compelled by insufficient wages to expose their wives to the degradation of field labour, and to send their children to work as soon as they could crawl, the labourers would have been more than human had they not risen in an insurrection which could only be quelled by force. They had already carried patience beyond the limit where it ceases to be a virtue.[47]

Poet Shelley, like Cobbett whom we have quoted at the beginning of this chapter, emphasized the difference between the past of the poor and their present; and he noticed the longer hours:

> Not that the poor have rigidly worked twenty hours, but that the worth of the labour of twenty hours, now, in food and clothing, is equivalent to the worth of ten hours then. And because twenty hours cannot, from the nature of the human frame, be exacted from those who performed ten, the aged and the sickly are compelled either to work or starve. Children who are exempted from labour are put in requisition, and the vigorous promise of the coming generation blighted by premature exertion. For fourteen hours' labour, which they do perforce, they receive – no matter in what nominal amount – the price of seven. They eat less bread, wear worse clothes, are more ignorant, immoral, miserable and desperate.[48]

Prothero and Shelley were describing the situation among the farm workers, who were still in the period the largest segment of the labouring population. Though this has not gone unnoticed, it has often been ignored in the belief that the essential test of betterment, looking into the future, was the industrial worker. What about his condition? For one category – the largest, in fact – the situation did not get better, but unmistakably worsened. By the 1830s, the handloom weavers had been reduced to a wage of less than a penny an hour: they were able to keep alive only when their children and wives joined them in the factories. The application of steam power to looms gradually undermined their independence and their number. They did not give up easily, but they had to in the end, provoking Ashton to term the period one of the most depressing chapters in the economic history of the time.

The factory system needs less comment: when factories first appeared the owners found it difficult and often impossible, in spite of what White or Landes say, to persuade men and women to work in them. Later, even if many adult workers found themselves in better-paid jobs than they could have hoped for if the factories had not been created, the tens of thousands of children working in a way recorded history has never known before, is a fact civilized men will never be able to exorcise from their minds.

To get back to the factory's bleak appeal, Inglis' sketch of the period captures it briefly and well:

> When Owen first went to New Lanark, for example, his predecessor there explained that he had been compelled against his will to use pauper apprentices because such was the dislike of factory work that, with few exceptions, "only persons destitute of friends, employment and character were found willing to try". If convicts had been compelled to work a twelve-hour day as part of their punishment, in jails, it would have provoked a humanitarian outcry. Yet the twelve-hour working day in factories had been established on commercial grounds – and not just as the norm; as the minimum. It was this, rather than the cruelty involved, which was the ugliest aspect of the factory system: that it imprisoned men, women and children for so much of their lives.[49]

The mill-owners did not deny the cruelty, they merely found the discussion about it irrelevant. In their opinion, the factory worker *was* better off. He was enjoying a standard of living higher than he otherwise could have hoped for, especially if his lot were compared to a century earlier when there had been no factories. If it were not for the efficiency and economies of factory production, leading to better trade, there would be no job for him to do and he would be faced with the alternative of the workhouse or starvation. The mill-owners were in a sense right, and the young Macaulay, who would later significantly, turn out to be one of England's most influential imperialist historians, felt justified in arguing that English labourers were no worse off than their counterparts on the Continent. The point, however, is whether *they ought not to have been better off*.

Therefore, when Karl Marx began to talk of surplus value, everybody understood what he meant: while theoreticians have reacted, often pathologically, to his prediction of the fall of capitalism, no writer worth his salt has felt it necessary to dispute the empirical content of surplus value itself, for it was really based on bare experience. In 1834, Jean de Sismondi, accepting the fact that machinery had vastly increased England's productive potential, and had made fortunes for many employers and enabled England to become the foremost trading nation in the world, still made it known that all of it had been built up only at the expense of the worker:

> The proletariat are cut off from all the benefits of civilization; their food, their dwellings, their clothes are insalubrious; no relaxation, no pleasures except occasional excesses, interrupt their monotonous labours; the introduction of the wonders of mechanics into the arts, far from abridging their hours of labour, has prolonged them; no time is left them for their own instruction or for the education of their children; no enjoyment is secure to them in those family ties which reflect their sufferings; it is almost wise for them to degrade and brutalize themselves to escape from the feeling of their misery; and that social order which threatens them with a worse condition for the future, is regarded by them as an enemy to combat and destroy. And this is not all: whilst their own distress is increasing, they see society overcome, as it were, by the weight of its material opulence; they are in want of everything; and on all sides their eyes are struck with what is everywhere superabounding.[550]

THE RISE OF A TECHNOLOGIZED CULTURE

The sort of pressures I have described in the larger part of this chapter brought about an awesome, lasting, and pervasive change in the structure of English society itself. The industrial revolution brought the majority of the population of England to face a situation in which the total adaptation of their lives to the rigours of a new productive system became a virtual necessity. By the time further technological changes arrived, a new generation had grown up, as David Landes observes, "inured to the discipline and precision of the mill". No wonder the workers were reluctant to enter the factory. A committee report of 1834 reported that "all persons working on the power looms are working there by force".

We are still not at the core of this issue. Eric Hobsbawm comes close when he writes that material poverty went hand-in-hand with social pauperization: the destruction of old ways of life without the substitution of anything the labouring poor could regard as a satisfactory equivalent.[51] The upper classes did not face this problem, and Hobsbawn has noted further that if there was indeed a relation between the industrial revolution as a provider of comforts and as a social transformation then those "classes whose lives were least transformed were also, normally, those which benefited most obviously in material terms . . . ."[52]

Thus, if we have the majority poor in mind then, on balance, it seems likely that English society had to pay for increased production of basic subsistence items by undergoing a worsening of cultural, social, and environmental conditions during at least a part of the nineteenth century. I have observed that people only accepted the rigours of industrial life in the hope of improving their subsistence; in the bargain they came close to suffering severe cultural, social, and environmental deprivation: they came near to living on bread alone.

Poverty in one sphere was exchanged for poverty in others that seemed less vital at first: entertainment, education, and social activity. And it is the lack of these forms of experience that have created the urgency of the consumer society. New needs sprang up because of the changed life-styles: the old methods of satisfying many human needs were destroyed or rendered obsolete. As Hobsbawm has put it:

> Pre-industrial traditions could not keep their heads above the inevitably rising level of industrial society. In Lancashire we can observe the ancient ways of spending holidays – the rush bearing, wrestling matches, cockfighting and bull baiting – dying out after 1840; and the forties also mark the end of an era when folksong remained the major musical idiom of industrial workers.[53]

Secondly, new needs have also sprung from the new pattern of living itself. Society has relied heavily on the economic system to right this situation. Man has had to be encapsulated increasingly in his own creations to make his urban life-style workable. Mass culture has been the logical result.[54] This is to be distinguished from what is today seen as a new phenomenon: a huge eruption of interest in, or curiosity about, (high) culture itself, on a scale unprecedented in human history and which seems to have caused aristocrats like Ortega y Gasset some anxiety. The economic system has, of course, been able to discover even the lucratic possibilities offered by this new urge, but how much this interest in culture has been deflected into an interest in what is provided under the name of culture will never be known.

There were undoubtedly many aspects of the pre-industrial way of life which were especially satisfactory, and it was only after the disruption of this way of life that people experienced some particularly pressing needs outside the sphere of traditional subsistence. Without the enormous increases in incomes and consumption which continued industrial development has produced, members of industrial societies would surely have been worse off than their agricultural predecessors. In human terms the real standard of living in the early and mid-nineteenth centry was abysmal. Incomes had to increase sufficiently to offset the losses before real progress was possible.

Homo ludens of rural festival life, of leisurely companionship, suffered an eclipse during the industrial revolution, but music halls, competitive factory brass bands, football clubs, tried to raise him up again. In the twentieth century, the cinema, radio, popular music, record players, and, above all, television, fulfil a similar function.

Both modern entertainment and the mass media indicate in their content and styles that they have grown up to fill a specifically social vacuum. A great deal of the programmes attempt to create the direct illusion of a close personal relationship with their audiences. Or the press. In a community where there is a high level of interaction between members, the community itself provides behavioural norms for its members – social approval, disapproval, rewards, and sanctions are generated from within. Today, the press plays an increasingly important role in this, and sometimes steps in as a reference point round which people might develop their attitudes and behaviour. Is it any more unusual that we can even identify a person's norms by the kind of newspapers he carries and reads?

The question is not whether, in the final analysis, pre-industrial or primitive societies would not enjoy some of the goods of modern life; rather one should ask whether pre-industrial populations would be prepared to work the long, tedious hours for these goods and services. Articles of consumption-oriented societies will not necessarily meet the personal or social needs of cultures that prefer to preserve their leisure. It should come as no surprise to discover that it was already in the mid-eighteenth century, as the industrial revolution gained momentum, that attitudes to leisure changed, and labourers began to use any rise in earnings to increase their consumption rather than their leisure time.

Writes Wilkinson:

> It appears that industrialization requires a more extravagant lifestyle than the modes of production that preceded it. The problems it creates and the needs it sets up make increased consumption a necessity if people are to lead reasonably satisfactory lives. The continuous expansion of gross national product which this requires should perhaps be regarded more as a reflection of the rising *real cost* of living than an indication of increasing welfare.[55]

It does appear from what has been said thus far that rich societies are less rich and poor societies less poor than has been hitherto imagined. I suggest that the words "developed" and "rich" be dropped when describing the industrial nations, and that the adjective "sick" might do a better job. And that we term the Southern countries, in so far as they continue to insist on aping the West, "confused".

CHAPTER FIVE

Technology, Culture, and Empire: The Colonial Age

> To some extent, the mechanical age seems linked to the age of colonialism. Both reached their apogees in the nineteenth century: both were based on the instinct for exploitation. The world was prospected to discover and cultivate raw materials with which to feed the machines. It rarely occurred to the ruling powers that the people who toiled to produce these materials, and who sweated to bring them forth, should have any appreciable use and benefit from other products. Whenever the natives made any serious trouble, the usual response was to send a gunboat.
>
> –K.G. Pontus Hulten: 1968

In 1498, Vasco da Gama opened the sea route to India, and simultaneously a new chapter in Asian history that would terminate a long four hundred and fifty years later with the withdrawal of British forces from India in 1947 and of the European navies from China in 1949.

It is important to note here that *before* 1498, the civilizations of Europe, India, and China had been virtually, and in a greatly limited sense, geographically isolated from one another: the fact of "technological osmosis" between these civilizations is the reason for the qualification. Even after 1498, in fact, till the year 1800 (as a mean), the relations between East and West still continued to be

> conducted within a framework and on terms established by the Asian nations. Except for those who lived in a few colonial footholds, the Europeans were all there on sufferance.[1]

Support for this assessment made by Donald Lach comes from another influential historian, G.B. Sansom, who wrote in 1949:

A survey of the enterprises of Europeans in Asia after the great voyages of discovery shows that during the sixteenth, seventeenth and eighteenth centuries neither their colonizing and trading activities nor their missionary work brought about any significant change in the life of the peoples with whom they came into contact. The presence in Asiatic countries of small groups of European officials and traders made little impression upon indigenous cultures outside a very narrow circle. . . . Indeed, far from Europe affecting Asia, it was Asiatic goods that changed and enriched European life, and Asiatic ideas that attracted some European minds.[2]

The Asian historian, K.M. Panikkar, concurs:

Essentially, till the nineteenth century . . . there was no large demand for European goods in any Asian country. The Empires of Asia . . . had, generally speaking, self-sufficient economies. Though the trade of India was large at all times, the economy of the country was not based on trade. This was true of China also, and the imperial government seems at all times to have discouraged the import of foreign goods into its territory.[3]

This state of affairs was recognized, and where not, soon made obvious in the opinions of men in China, India, and Britain during the period. The edict of Ch'ien-lung, who in the eighteenth century (1793) received an embassy from King George III of England, is worth quoting even in part:

The various articles presented by you, O King, this time are accepted by my special order to the officer in charge of such functions in consideration of the offerings having come from a long distance with sincere good wishes. As a matter of fact, the virtue and prestige of the Celestial Dynasty having spread far and wide, the kings of the myriad nations come by land and sea with all sorts of precious things. Consequently there is nothing we lack, as your principal envoy and others have themselves observed. We have never set much store on strange or ingenious objects, nor do we need any of your country's manufactures.[4]

For the two hundred and thirty years after Albuquerque's disastrous attempt to challenge the power of the Zamorin of Calicut (1506) – he had to be carried unconscious to his ship – no European nation attempted any military conquest or tried to bring any ruler under control. In 1739, for example, the Dutch who came up against the Raja of Travancore had to surrender. A year earlier, the British naval authorities on the West coast reported:

Our strength is not sufficient to withstanding him [Sambhaji Angria] for I assure Your Honour that he is a stronger enemy than you and a great many others think.[5]

We should also remember that a Company settlement was made possible in Madras in 1708 only after a grant of five villages was made by the regime in Delhi for the purpose; and that the Viceroy of Bengal continued to be addressed in the most cringing terms. In addressing the Emperor one of the Englishmen described himself as

the smallest particle of sand, John Russel, President of the East India Company with his forehead at command rubbed on the ground.[6]

Europe at the time had but little to offer to the Asian economies. Illuminating in this regard is the attempt of the English to sell textile goods to the East. The production of woollen goods in England in the sixteenth century was always in excess of home consumption and the export of woollen cloth was vital to the English economy. However, two wars on the Continent – the Spanish War and the French Civil War – soon put this market into jeopardy.

A serious crisis of overproduction in the 1550's stressed the need for new outlets and the most hopeful prospect then seemed to be that of establishing trade with the Far East: both China and Japan, it will be noted, have cold climates. The voyages of 1554 (Willoughly and Chancellor), 1575 (Frobisher), 1578-83 (Drake), 1585 (Davis), 1596 (Wood), and, finally, the founding of the East India Company in 1600, all these were inspired, directly or indirectly, by England's pressing need to sell its own textiles in the Far East. Sansom admits this when he writes that "the history of English commercial enterprise in the Far East shows a continuous but abortive effort to find a 'vent' for English woollens".[7]

Of course, there was another objective too, which was to compete with the Portuguese and the Dutch for the pepper and spices of the Malay Archipelago, but the principal aim remained the selling of *English* textiles in Asia. History books perhaps generally ignore this, because it proved to be so total and embarrassing a failure.

The Company's attempts to establish trade with China were unsuccessful. So it next tried to dispose its English woollen cloth on the spice-islanders. Here it discovered that the only commodity acceptable was *Indian* textiles and this prompted it to seek a market for its woollen goods in India, with the idea of buying in return the Indian cottons and silks wanted by the spice-islands. With this end in view, English ships reached Surat (Gujerat, India) in 1608. Here, again, the Englishmen tasted failure. And three years later, the Company's factors wrote to the London Directors:

> Concerning cloth, which is the main staple commodity of our land . . . it is so little regarded by the people of this country that they use it but seldom.[8]

A decade later, the Company had finally abandoned hope of a big Asian market for English broadcloth. Yet, some other commodity had to be bartered if the Company wished to get its hands on the spices and pepper of Malay. Other alternatives were tried, including consignments of looking-glasses, sword-blades, oil-paintings, drinking glasses, quicksilver, coral, and lead. To stimulate the demand for English lead in India, it was decided to send out "plumbers to teach them the use of pumps for their gardens and spouts on their houses".

This was followed by a scheme to persuade Jahangir (the Moghul Emperor) to pay for the erection of waterworks for the supply of Agra, knowing that it would require plenty of lead piping. There is another story of how the London Directors, hearing that Indians "are very superstitious and washe their hands whensoever they goe to their worship", immediately ordered the despatch of a consignment of wash-basins for trial sale. But to no avail. In the end, the factors were forced to conclude that "no commodity brought out is staple enough to provide (in return) cargo for one ship".

Earlier, the Portuguese had faced a similarly disheartening situation. The presents offered by da Gama to the King of Calicut, included some striped cloth, hats, strings of coral beads, wash-basins, and jars of oil and honey. As Sansom observes, these were curious gifts to bring to the classical land of treasure, and the King's officers readily found them laughable. In fact, when he received da Gama, the King had in his hand a golden spittoon and by him a golden basin for his betel. He was wearing a crown set with pearls, and golden anklets set with rubies.[9]

Panikkar continues, with some more amusing evidence:

> The story of the Amsterdam Company which exported to Siam a collection of thousands of engravings, of madonnas and biblical scenes, "prints recording the stories of Livy and, finally, prints with a more general human appeal, a collection of nudes and less decent illustrations" is not by any means strange or unique. Richard Cocke's letter from Japan complaining of the lack of interest in Biblical paintings may also be quoted here. "They es-

teem a painted sheet of paper with a horse, ship or a bird more than they do such a rich picture. Neither will any give six pence for that fair picture of the conversion of St. Paul."[10]

To return to the Company, it was compelled to fall back on the export of bullion (in the form of gold and silver) for the purchase of goods in India. This raised a storm in England from people who feared that the country was being drained of its wealth. The Directors were able to counter this by the re-export of Indian goods to the European Continent and the Levant, thus recovering the return of the original wealth, plus a trade profit. Later, they were able to pay for Chinese goods with Indian opium.

The self-sufficient economies of Asian countries before the arrival of the Europeans had in fact stimulated between themselves a flourishing trade in the form of semi-luxury and luxury goods. C.G.F. Simkin has amply documented and detailed the various trade routes that had long been established between Japan, Korea, China, East Turkestan, Tibet, north and south India, Ceylon, Thailand, Laos, Cambodia, Vietnam, Sumatra, Afghanistan, Iran, Java, Malaya, Burma, the Philippines, and Taiwan. And the history of European expansion in Asian waters in nothing more than the usurpation of this trade, through means more foul than fair.

Merely the example of Malacca in the fifteenth century is sufficient to give some idea of the extensive nature of the "traditional trade" of Asia as a whole. Four groups of harbour masters (*shahbandars*) were responsible for the four different points of trade: one looked after ships from Java and the other parts of Indonesia; a second met ships from China, Champa, Burma, and Thailand; a third, ships from Bengal, Malabar, and North Sumatra; and a fourth, principally ships from Gujerat. Such was the diverse nature of merchants that gathered there that a special Malay *lingua franca* soon evolved to meet the confusion of eighty different languages or dialects.

The trade between Malacca and Gujerat was described by Tome Pires, a Portuguese official who visited the former trading centre between 1512 and 1515. He wrote:

Four ships come every year from Gujerat to Malacca. The merchandise of each ship is worth fifteen, twenty or thirty thousand cruzados, nothing less than fifteen thousand. And from the city of Cambay one ship comes every year; this is worth seventy or eighty thousand cruzados, without any doubt. The merchandise they bring is cloth of thirty kinds, which are of value in these parts; they also bring pachak, which is a root like rampion, and catechu, which looks like earth; they bring rosewater and opium; from Cambay and Aden they bring seeds, grains, tapestries and much incense; they bring forty kinds of merchandise....[11]

The fact is the Portuguese were able to build their sea-borne empire largely from this trade *within* Asia. Sansom has observed that it was the profits earned by the Portuguese as carriers and brokers that sustained, for example, their commerce with China, rather than the sale of the few European products for which they could find a market.[12] John Harrison has summed up the matter as follows:

The Portuguese used their naval power to draw Asian shipping to their ports and customs-houses – and to plunder those who did not give either submission or a timely bribe. Officials and settlers themselves engaged, alone or in partnership with Asian merchants in the various trade systems of the Indies. This "country trade" was far greater than that with Europe. One complex system interchanged the goods of East Africa, the Red Sea and Persian Gulf and India; this was linked by Indian cottons and opium to the nutmegs and cloves, tin, copper and gold, porcelains and silks assembled at Malacca, which drew in turn about the trade between Siam, China, Japan and the Spice Islands. The Portuguese helped to link all these together, and in the process earned the means to defend Malacca or to embellish golden Goa.[13]

The strength of this largely autonomous (in relation to Europe) Asian trade was bound to have its consequences on the economies of Europe once these were plugged into the current. "One is apt to think of European intrusion," wrote Sansom, "as producing disturbances in Asiatic life and to forget that from their adventures European countries experienced effects which were not all beneficial."

The import of Indian textiles into England threatened to upset the woollen trade there. As early as 1695, in fact, Parliament was urged to prohibit the use of Indian fabrics, as the visible depression and unemployment in the English woollen trade was no longer possible to ignore. The demonstrations raised by the woollen and silk weavers alarmed the government, and in 1700 there followed the bill prohibiting the home consumption of silks or printed calicoes of Asiatic origin. To quote Sansom again:

The Asiatic trade not only changed our mode of dress and introduced new aesthetic principles, but also altered in the long run the constitution of English commerce and even the trend of English economic thought. It was the controversy over calicoes that presented in an acute form to people and Parliament the choice between protection and free trade and ultimately – after a protectionist phase – led to the adoption of those doctrines of free enterprise and *laissez-faire* which dominated English theory and practice for many generations.[14]

Also of equal significance was the nature of the encounter of Europe, China, and India on the level of their cultural systems. Panikkar has a comprehensive chapter on the attempts to Christianize the East and their ultimate failure: the Jesuit influence in China has been vastly exaggerated in the past. The fact is the Chinese authorities never did fail to keep the Jesuits and their mission under control. Further, the scientific contributions of the priests, when placed in proper perspective, are truly ambiguous.

We must remember that the work of the astrological bureau (euphemistically termed "astronomical" by the Jesuits), and to which Jesuit Adam Schall von Bell was appointed, was not strictly speaking a scientific one. Its principal business included the preparation of an official calendar, containing auspicious dates for almost every event of domestic or national importance. This *compromise* to which Ricci and Schall were forced was noticed by fellow Jesuits who reacted to it in the strongest terms. As Panikkar has noted, Schall was being plainly dishonest and unscrupulous when he did not hesitate to interpret the sun's spots on one occasion as representing the hostile influences of the Buddhist priests near the Emperor. Writes the Asian historian: "Schall and his friends were supposed to be promoting the truth of their religion by this kind of deceit."

The favourite strategy of the Jesuits, from Ricci onwards, had involved a continual critique of Buddhism in order to raise themselves in the eyes of the Confucian literati. The permanent patronizing attitude of the Chinese to the entire Christian mission is obvious. Later, during a particularly weak Chinese phase, the re-entry of the Church hand-in-hand with imperialism (the Bible in one hand and opium in the other) would prepare the way for the massacres of the Boxer uprising.

Today, the West has replaced Christianity as the new ideology inextricably connected with imperialism. Ultimately, it might have to suffer the same pattern of events that Christianity had to face in India and China. The re-establishment of Asian trade, the renewal of Asian cultural systems, the re-assumption of political autonomy by Southern nations for the determination of their own technological futures might conceivably lead to a situation that existed before the European impact and which we have briefly outlined above. The relativization of the Western paradigm is not inevitable. But it is desirable and necessary in the light of the attempts by the nations of Asia, Africa, and Latin America to emancipate themselves. The eventual success of these attempts would make the relativization inevitable.

INDUSTRIALIZATION AND DE-INDUSTRIALIZATION

After 1500, civilizations as a rule do not rise, fall, and decline in isolation, but interconnected with others, begin to follow the movements of a see-saw. For 300 years after that date, Europe seems to have been on the defensive. For the past 175 years to the present day, Asia has seen fit to lie low. It is instructive to follow this 175-year-old history of colonialism and independence.

Not that we are still quite clear about what happened in those years: there have been so many revisions. Simply put, we just do not quite understand the economic history of the period, particularly in relation to China and India: there are too many loose ends. Central to the entire discussion is the claim of the colonial nations that the industrialization of the Northern nations presupposed and perpetuated the de-industrialization of their Southern counterparts. Let me begin with India and present the de-industrialization case.

The Moghul Empire declined in the first half of the eighteenth century: more precisely, effective central control over the Empire's territories was loosened and lost after the death of Bahadur Shah I in 1712. However, an all-important point needs immediate stressing here: the absence of central political domination did not discourage, neither did it disintegrate material culture, a phenomenon not at all unique in Indian history.[16]

That the decline of central Moghul power did not mean much to the economy is evident from a quick look at the trade figures of the economy *after* Moghul decline. In 1708, Britain imported goods from India worth £493,257, and exported in return goods worth £168,357. By 1730, while the imports to England rose to £1,059,759, the exports fell to £135,484. In 1748, imports into Britain were still £1,098,712 and the exports had declined further to £27,224. The balance was paid by Britain in bullion; in fact, between 1710 and 1745, India received £17,047,173 in bullion.[17]

By 1757, the East India Company, with the support of a powerful clan of Hindu merchant capitalists, had gained a foothold in the politics of Bengal. The Hindu merchants were keen to associate with foreigners, since they controlled the trade and could thus reap huge profits – a situation not much changed today when Indian capitalists welcome multinationals. The East India Company received the right to the revenue of a district: the twenty-four *pargannahs*. By 1764, the Moghul Emperor was forced to extend the revenue rights of the Company to other territories in Bengal, Bihar, and Orissa.

The Company's early administration in Bengal is too sordid to be repeated here in detail: it used its monopoly position to impose taxes of numerous kinds on different products including salt, betel-nut, tobacco.[18] The effects of Company rule on the textile industry are better known.

The Indian textile industry declined *before* the industrial revolution in Britain. The displacement of the Muslim aristocracy simultaneously displaced domestic demand. A famine in 1770 may have reduced the population of Bengal by a third. Equally deleterious was the conduct of the Company towards the weaver. Total political power allowed its men to ensure that the entire produce of the area was sold to them. As a document of the time noted:

> They trade . . . in all kinds of grain, linen and whatever other commodities are provided in the country. In order to purchase these articles, they force their money on the ryots and having by these oppressive methods bought the goods at a low rate, they oblige the inhabitants and the shopkeepers to take them at a high price, exceeding what is paid in the markets. *There is now scarce anything left in the country.*[19]

After the Company took over the administration of Bengal, the once-favourable balance of trade was reversed. Already in 1773, a report made to Parliament calculated revenue collections to be £13,066,761 for the six years of possession: expenditure amounted to £9,027,609, and the balance with the Company £4,037,152. The Company, in fact, now had a revenue surplus larger than the Indian surplus on the commodity trade with Britain. The Company surplus was used to purchase Indian products for export into Britain: thus did the colonial "drain" begin.[20]

Maddison notes the intricacies involved in repatriating the balance:

> In order to effect the transfer of these additional resources, some Indian bullion and diamonds were shipped to the U.K. and Bengal silver was

exported to China to finance British purchases of Chinese tea. In addition, Company servants sold their rupee profits to foreign trading companies against European bills of exchange, which supplanted other countries' exports of bullion to India. Bengal had a surplus on trade with other parts of India and these revenues were used by the East India Company to finance military campaigns in Madras and Bombay. Bengal revenues and profits were also used to finance the local costs of a larger contingent of Company servants and private traders. The annual net real transfer of resources to the U.K. amounted to about £1.8 million a year in the 1780s. This was also the size of India's exports.[21]

Indian cotton manufactures continued to be imported into Britain. In fact, they reached their peak in 1798, and even in 1813, they still amounted to a value of £2 million. The industrial revolution in Britain had already revolutionized the making of textiles; between 1779 and 1812, in fact, the cost of making cotton yarn dropped nine-tenths. Why, then, did Indian goods still hold their demand?

Because even thirty years after the industrial revolution transformed textile production, Indian textiles still remained *cheaper* than machine-made goods: this can probably be explained by the fact that the weaving processes in England had not yet been extensively mechanized. About the relative cheapness of Indian goods, the historian H.H. Wilson has this to say:

It was stated in evidence (in 1813) that the cotton and silk goods of India up to the period could be sold for a profit in the British market at a price from 50 to 60 per cent *lower* than those fabricated in England. It consequently became necessary to protect the latter by duties of 70 and 80 per cent on their value, or by positive prohibition. Had this not been the case, had not such prohibitory duties and decrees existed, the mills of Paisley and Manchester would have been stopped in their outset, and could scarcely have been again set in motion even by the power of steam . . . the foreign manufacturer employed the arm of political injustice to keep down and ultimately strangle a competitor with whom he could not have contended on equal terms.[22]

Confirmation about this comes from another independent source, the German economist, Friedrich List, who wrote the following in 1844, in his volume *The National System of Political Economy:*

Had they sanctioned the free importation of Indian cotton and silk goods into England, the English cotton and silk manufactories must, of necessary, soon come to a standstill. India had not only the advantage of cheaper labour and raw material, but also the experience, the skill, and the practice of centuries. The effect of these advantages could not fail to tell under a system of free competition. . . .[23]

The high tariffs were exclusively raised to make Indian textile goods more expensive than the products of the machine. Wilson was referring to the following passage that constituted the evidence of John Ranking, a merchant, examined by the Commons Committee in 1813·

Can you state what is the ad valorem duty on (Indian) piece goods sold at the East India House?

The duty on the class called calicoes is £3.6s.8d per cent upon importation, and if they are used for home consumption, there is a further duty of £68,6s,8d per cent.

In this session of Parliament there has been a new duty of 20 per cent on the consolidated duties, which will make the duties on calicoes. . . used for home consumption, £78.6s.8d per cent, upon the muslins for home consumption, £31.6s.8d. per cent.[24]

It is important to stress that this tariff system was primarily intended to protect Britain's infant factory system even in the first two decades of the *nineteenth* century. I have already indicated that the weakness of the English industry probably lay in the fact that the weaving industry had not yet been extensively mechanized. Though the power-loom was invented by Cartwright in 1787, its adoption, as David Landes writes, "was slow during the first two decades of the century". The glut of machine-produced yarn made the situation in fact "a golden age of the hand-weaver, whose unprecedented prosperity was a shock to all, scandal to some."[25]

In sum, machine-produced yarn might have been vastly cheap,[26] but the cloth made from it, hand-woven, still proved more expensive than Indian hand-woven textiles in Britain: it therefore became necessary at this stage for political power to intervene and support the machine. Also at this time a new circumstance on the Continent reinforced the new sequence of events.

What had happened with the woollen trade earlier now began to affect the newly mechanized English textile industry in cotton. The Napoleanic wars excluded British manufactures from the Continental ports and English merchants and manufacturers began to feel the need for fresh areas to "vent" their cotton goods. Up to 1813, the East

India Company had been allowed the monopoly of trade with India, and this trade was primarily devoted to the export of Indian manufactures. In 1813, however, the House of Commons held hearings on the Company's right, which resulted eventually in the abrogation of the monopoly: the trade now passed into the hands of private merchants financed by East India Agency Houses. The stage was set for the large-scale dumping of English goods on the Indian market.

The first to be affected by the rise in the imports of English cotton goods into India were Indian female spinners. In 1828, for example, cotton yarn and twist imported into India through Calcutta already totalled 1.2 millib lbs; this volume was further increased to 3.2 million lbs in 1835 and to 17.5 million lbs in 1847. That there was very little chance of competing with machine imported yarn is indicated by the following computations of 1840:

Prices of 1½ hanks:

	English			*Indian*		
Count of yarn	Rs.	As.	gds.	Rs.	As.	gds.
200	0	3	00	0	13	00
190	0	2	15	0	10	00
180	0	2	15	0	6	00
170	0	2	10	0	5	10
160	0	2	10	0	4	00
150	0	2	10	0	3	10

(Source: A Pearse: *The Cotton Industry of India* 1930.)

In Britain, on the other hand, the power-loom was being used on a wider scale after 1815. David Landes notes that there were already 2,400 of them in 1813, and that the number had risen to 14,150 in 1820, to 55,000 in 1829, to 100,000 in 1833, and by mid-century had crossed 250,000. In 1814, the quantity of cotton goods exported to India from Britain had been a mere 818, 208 yards; in 1835, the figure had risen to 51,777,277 yards.[27]

Duties on Indian goods imported into Britain were finally repealed in 1846, when Britain legally accepted the *laissez-faire* ideology. By then, of course, the British factory system's foundations had been firmly cemented. There still remained the problem of silk: fine silks could not be woven by power. Yet a great deal of raw silk had been continuously imported into Britain in the 1820s, where it was worked and later exported to European markets. The French introduced a new element into the picture.

Till the thirties, British silk goods had done well in France, where Indian goods were officially prohibited. Once the prohibition was removed, the entire British trade collapsed in favour of Indian silks. The export of raw silk from India began to decline: in 1829, India had exported silk worth £920,000. By 1831, this raw silk export had fallen to £540,000: more raw silk was being used in India for manufactures for export. In 1832, British silk exports to France had been valued in the region of £50,00 and India's at £29,000. By 1839, the British contribution had shrunk to £5,500 and India's stood at £168,500.

The duty on Indian finished silk goods into Britain was fixed at 20 per cent, while British finished silk goods to India paid a nominal duty of about 3½ per cent. A proposal to equalize the duties was rejected by a Select Committee, to protect British labourers. The following discussion between Mr. Brocklehurst, a representative of British industry and Mr. Cope, a silk weaver in Britain, is not only significant, but has contemporary connotations too:

Mr. Brocklehurst: What would be the effect upon this branch of your trade if the present duty on East Indian silk goods were reduced from 20 to 3½ per cent?

Mr. Cope: In my opinion, it would have the effect of destroying this branch of trade; and if so it would rob of their employment, and consequently of the means of living honestly by their labour, all those parties which I have named, and would make them destitute and reckless, and cause them to become a burden to the rest of society, whose burdens are already too heavy. It would throw out of employment a large amount of capital and would give into the hands of foreigners that employment by which we ought to be supported.

Mr. Elliott: Do you think that a labourer in this country who is able to obtain better food. . . has a right to say, we will keep the labourer in the East Indies in that position in which he shall be able to get nothing for his food but rice?

Mr. Cope: I certainly pity the East Indian labourer, but at the same time I have a greater feeling for my own family than for the East Indian labourer's family; I think it is wrong to sacrifice the comforts of my family for the sake of the East Indian labourer because his condition happens to be worse than mine; and I think it is not good legislation to take away our labour and to give it to the East Indian because his condition is worse than ours.[28]

There is a clear pattern in the attempts by British manufacturers to convert India after 1813 into a complementary satellite economy providing raw materials and food for Britain and an ever widening market for its manufactures. All this could be accomplished through the means of a practical economic imperialism, in which the arts of political manipulation gave aid to the craft of enterprise and British technology and in which the dominion employed by the superpower came to be associated with wilful and effective subordination.

Contrary to the views of Lenin and Hobson, imperialism was not a late-nineteenth century phenomenon that rose with the decline of free-trade beliefs. It was required earlier to support the rise and strengthening of the machine. Twenty years after the enshrining of the free trade legacy, Richard Cobden, one of the chief pillars of the Manchester school suggested that the principles of Adam Smith did not govern relations between Great Britain and India.[29] A year before that, in 1862, Thomas Bazley, the President of the Manchester Chamber of Commerce, had already decided that the "great interest of India was to be agricultural rather than manufacturing and mechanical".

The free traders with their *laissez-faire* attitudes were irked beyond reason by those nominal duties the Indian colonial government levied on English imports into India. As Harnetty notes:

> The full development of India as a source of agricultural raw materials (and this meant, of course, cotton) was inhibited by the Indian cotton duties which, by protecting native manufactures, caused the consumption in India of large quantities of raw cotton that otherwise, i.e., under "free competition", would be exported to Great Britain. It followed that the duties must be abolished, thereby enhancing the supply of cotton for British industry and enlarging the market in India for British manufacturing goods. Such a policy could be justified on theoretical grounds by the doctrine of free trade.
>
> But to encourage India as a producer of raw materials required more than economic freedom. It also involved a contradictory policy of governmental paternalism. Lancashire demanded that the Government of India inspire the development of private enterprise in the Indian empire by financing some of this development. In line with this demand, the authorities in India guaranteed railway construction and undertook numerous public works. They also undertook the experimental cultivation of cotton and, in this connection, made the first attempt at state interference in India in the fields of production, marketing and trade.[30]

In 1860, the East India and China Association was still protesting that a new increase in the cotton duties in India (necessitated by a deficit in the Indian budget) would give a "false and impolitic stimulus to yarn spun in India, thereby serving to keep alive the ultimately unsuccessful contest of manual power against steam machinery". Another petition from the Manchester Chamber of Commerce in 1860 could continue to claim that any new tariff on British imports into India would harm not only the manufacturers of Great Britain but also the population of India "by diverting their industry from agricultural pursuits into much less productive channels under the stimulus of a false system of protection".

In the same year, the Board of Trade was supporting the case of English bleachers against the new tariff and noted that "in providing for a temporary emergency, a permanent injury be not inflicted on an important branch of the manufactures of these countries". Sir Charles Trevelyan, Finance Minister in India in the 1860s, was anxious to see the disappearance of the Indian weaver as a class, a development he thought best for both Britain and India: India would benefit because the weaver, faced with competition from machine-made goods, would be forced to give up his craft and turn to agriculture; the increased labour supply would then raise output and England would benefit since makers of cloth would be converted into consumers of Lancashire goods.[31]

It comes as no surprise to learn that when the cotton duties were totally abolished in 1882, the Viceroy of India at the time, Lord Ripon, was privately willing to admit that it was political pressure rather than fiscal arguments which had led to their general repeal, and that India had been sacrified on the altar of Manchester.[32]

Peter Harnetty's excellent study, which we have used for some of the information discussed above, is valuable in that it sets out, with documented evidence, the almost yearly pressures and influence exercised by Lancashire interests on the Indian economy. This one-sided influence did not often go totally unchallenged: there were British men in India who tried, albeit unsucessfully, to fight against the stream. Such a characteristic personality was J.P. Grant, brave to make out his dissent:

> We think it our duty to submit our earnest protest against the principle that the taxation of India is to be regulated under pressure from powerful classes in England, whose private interest may not be the interest of India, and with regard to the principle established in England and for England, and without ascertaining by communication with the responsible Government in India the

policy or financial bearing of the measure or the views and sentiments of our Indian subjects.[33]

Harnetty also demonstrates conclusively that the irrigation works, road building, railway construction, and improvement of inland waterways were undertaken at the instigation of Lancashire cotton interests, to ensure a dependable secondary source of cotton, especially in the first half of the sixties when the American Civil War disrupted nearly all supplies. The colonial government of India fell in with the schemes because it realized that any fall in the manufacturing capacity of Britain due to lack of raw cotton supplies would have been calamitous. Thus, the Governor of Bombay justified the exorbitant and wasteful financing of a nearly impossible feeder road between Dharwar and Karwar on the Indian west coast in 1862 in the following words:

The money value to India is very great, but its value to England cannot be told in money, and every additional thousand bales which we can get down to the sea coast before the season closes in June 1863 may not only save a score of weavers from starvation or crime but may play an important part in ensuring peace and prosperity to the manufacturing districts of more than one country in Europe.[34]

Likewise did a Chief Commissioner of the Central Provinces argue that construction of a railway would not only secure the more rapid export of raw cotton but also would lower the cost of imported Lancashire piece goods. This in turn would divert, he went on to observe, labour from spinning and weaving to agriculture and so lead to an extension of the areas under cultivation.

The de-industrialization thesis, of course, is not restricted to textiles: the privileged entry of a variety of other imports such as iron, paper, and glass soon dis-established Indian craftsmen involved for centuries in their fabrication. The iron industry may be briefly surveyed in this connection.

Dharampal has estimated that there might have been around 10,000 small furnaces in use in India in the early 1700s. By 1850, however, most of these furnaces had been knocked out of business. Vera Anstey rooted the decline in the indigenous industry in the massive imports of metal goods from Britain.[35]

It could be said almost immediately, that whatever might have been the condition of the average peasant or farmer in India during the period, there was little done by the British colonial government to improve it. But then, neither should anyone expect any colonial government to act positively in the first place. The colonial government in India was interested exclusively with arrangements that would guarantee their revenues from the land year after year, famine or no famine. The absence of such concern for the improvement of Indian agriculture did not in anyway lead to its deterioration: I have already given evidence earlier that the Indian farmer was quite expert in his work and needed little advice. British experts sent out to advise him were shrewd enough to recognize this sooner or later.

The British, for their part, took great pride in their ability to collect revenues even in the most difficult times. As an example: a serious famine hit Bengal in 1770, decreased the population by a third, and returned a corresponding third of the land to waste. A year later, Governor-General Warren Hastings could still write to the Court of Directors:

> Notwithstanding the loss of at least one-third of the inhabitants of the province, and the consequent decrease of the cultivation, the net collections of the year 1771 exceeded even those of 1768.... It was naturally to be expected that the diminution of the revenue should have kept an equal pace with the other consequences of so great a calamity. That it did not, was owing to its being violently kept up to its former standard.[36]

More important, the colonial government set about making institutional changes in agriculture by transforming traditionally restricted property rights into something more closely resembling the unencumbered private property characteristic of Western agricultural systems. William Woodruff sees in this one of the principal ideas that signalled the application of the Western idea of progress in the non-Western world. He even seems to regret that "over most parts of Africa, despite the intrusion of Western nations, the indigenous customs and laws which govern the division, ownership, and use of land and natural resources remain" and "provide a bar to modern enterprise".[37] The trouble with such theorizing is that it ignores the original purpose of these institutional changes (besides, of course, the unsettled question of property rights walking always hand-in-hand with progress): the new arrangements were made because they were found to yield, year after year, a guaranteed income to the British authorities.

Further, it was precisely this institution of property rights that ultimately turned out to be the principal "bar to modern enterprise". For, the consequence of this half-Westernized land policy, this change from custom to contract, was the creation of one of the greatest curses ever to settle on the structure of the Indian rural economy: the rise of the power of the moneylender. With the emergence of clear titles, it now became possible to mortgage land. I am not blaming anybody for this; it is only that I wish to point out that institutions conducive to better material welfare in one country may have totally unexpected effects in another country, especially when the latter finds itself ruled by an alien government.

Before the arrival of the colonizer, for centuries in fact, the moneylender had been nothing more than a servile adjunct to the cultivator, socially despised as much for his trade as for his religion. He was forbidden to wear a turban, permitted to ride only on a donkey, and often, as M.L. Darling observed, the object of "unmentionable indignities". It was British rule that freed him from restraint and armed him with the power of law, till he turned as oppressive as he had been hitherto submissive.[38]

The institution of property rights was specifically intended for the easier collection of the land tax: it established a direct legal relation between the colonial government and the peasant or landowner. This in itself led to the beginning of inheritance and thus to the problem of the subdivision of land. About the tax, though the amount was fairly heavy in the early years, the important point was its *rigidity*. Whatever the nature of the harvest, whether it was affected by drought or rain, a specific payment had to be made twice a year. As Mamdani writes:

> Given the fragmentation of land, one small drought was enough to drive an average peasant into debt. Not only that, but the tax had to be paid in cash. To obtain money for his crops, the peasant was obliged to sell them to the grain-dealer – who was also the moneylender.[39]

In the Punjab, for example, with the famines of 1860-61 and 1869, and the heavy mortality among cattle, the rule of the moneylender was firmly established. Moneylending became for the first time in Indian history the most profitable occupation in the area. See the figures. The

1868 Punjab census listed 53,263 bankers and moneylenders in the province. By 1911, there were 193,890, that is, a ratio of 1:100.

The influence of the moneylender on agriculture was a little less than disastrous. He had no desire to own land, for proprietary rights yielded little profit. If he purchased the land, he got merely the land, but as a mortgage he got the land and a hard-working, submissive *owner-tenant* as well. In 1875-76, 44 per cent of the cultivated land in the Punjab, for example, was held by farmers *who had become tenants;* and by 1919, the figure had risen to over 51 per cent.

Mamdani writes:

It should come as no surprise that under these circumstances the peasants – and particularly the owner-tenants who formed the majority of the cultivators – had little motivation for improving their agricultural methods and raising the productivity of the land. As agricultural prices increased, the increasing power of the moneylender was reflected in the gradual change from cash rents to rents in kind. If the peasant were to improve his farming methods and increase his productivity, the moneylender would simply demand a larger share of the produce.[40]

H. Calvert, a Registrar of Cooperative Societies in the Punjab in 1920, made this very obvious:

These tenants generally take less care in preparing the land for crops, plough it less often, manure it less and use fewer implements upon it than *owners*. They grow less valuable crops, especially avoiding those requiring the sinking of capital in the land; they make little or no effort at improving their fields; they keep a lower type of cattle; they avoid perennials and bestow no care on trees.[41]

Observe now how such situations have been wholly misunderstood and then distorted by writers like William Woodruff:

The idea of using money to improve the economy was largely alien to the Indian mind. In comparison with traditional investments in property, money-lending, trade, and jewellery, money spent in improving agriculture and industry bore a lower yield. Even where they bore a higher yield, ignorance and apathy had to be overcome.[42]

Similar platitudes have been proposed by Angus Maddison: he is convinced that the passivity of village life and the caste system had inhibiting influences on agricultural productivity. And David McClelland went as far as to suggest that training courses should be introduced in rural societies to breed an entrepreneurial class. What these theoreticians did not realize is that the behaviour of these peasants was most rational, having discovered for themselves that however beneficial any form of technology may be in the abstract, in actual fact, the benefits went to the moneylender, not to them. To outsiders, the entire system did indeed seem irrational, that is, inimical to the interests of the majority of the people in the village.

To return to the colonial government, the British did not merely change the structure of land-ownership relative to the rural economy, they set about interfering directly in the kind of commodity the farmer should produce. Nowhere is this clearer than in the production of opium for the Chinese market. The East India Company began to ship the narcotic to Canton first in 1773. After the Chinese imperial ban in 1796, the Company denied that it was further involved in the trade, but private traders plying the India-China route were licensed by the Company on condition that they carry only Company opium.

It was Warren Hastings again who first introduced a monopoly on the production and trade of the drug in India. All sorts of means, such as withdrawing tenancies or manipulating tenants into debt, were found to compel Bengali peasants to grow poppy and nothing else, not even vegetables for their own use.[43] The profits (2,000 per cent) were excellent and the direct importance of the trade to the British economy has never been denied. In 1801, for example, Britain spent £3.6 million on Chinese tea and since China did not need English goods, nine-tenths of the tea money was paid by the British in bullion. Opium reversed this trend.

Between 1821 and 1830, expanded production in India led to a jump in imports at Canton from 580,000 lbs to 2,913,000 lbs; the entire enterprise was supported by the strange notion that it would aid Christianity: the German missionary, Karl Gutzlaff, noted that the "traffic would tend ultimately to the introduction of the gospel". And another missionary, this time American, spoke of opium and naval forces as the instruments of "The Divine Will".

Back in India, even famine was no excuse: the shortage of food may have been terrible, yet several of the poorest farmers were compelled to plough up the fields they had sown with gram, in order to plant them with poppies. Contracts were forced on peasants for stipulated supplies, and a fine of Rs.300 levied on every chest short. The demand in India was first encouraged and consumption spread, until the Company realized that the addiction to opium considerably reduced the efficiency of native labour. Accordingly, the price was raised to discourage consumption, except for medicinal purposes and the surplus exported to China, where the deterioration in the condition of native labour would be of no loss to the Company. By 1870, half of China's imports consisted of opium and in India, next to land revenue, opium was soon the most productive source of colonial income. The burden fell naturally on the peasant:

> Great persecution is employed by the swarms of the peons to compel the ryot to take advances, and to devote a portion of his land to opium. . . . I have possessed extensive properties in the opium-cultivating districts; and I have seen ryots through tyranny, and to save themselves from persecution, compelled to sow opium in land belonging to me, even in the very compound of my house, which I have given them for other purposes.[44]

Activities related to agriculture also underwent changes due to British intervention: natural dyes, for example, were displaced by synthetic dyes (here the influence of Germany, however, was paramount). The imports of kerosene oil decreased the demand for vegetable oils, thus affecting the cottage oil-producing units. Raw jute, oil-seeds, raw cotton, and raw hides were now transferred to Britain in bulk as their processing and final manufacture became English privileges. The export of the seeds rather than the oil itself had consequences for Indian agriculture, as the *Review of Trade in India* pointed out already in 1879.

> It seems strange that the wasteful practice of exporting the seed should continue. It causes great pecuniary loss by waste and damage of the seed in transit from the place of production in the inland districts of India to the place of manufacture in France, England and the United States. The unnecessary expenditure in freight is a serious consideration and lastly it should not be forgotten that under the present system India literally throws away enormous quantities of oilcake, that is, an invaluable food for cattle and fertilizer for land. It is really a national misfortune that India should send away all this oil-producing material in the crude condition instead of pressing the oil in the country.[45]

I shall rest here with this parade: I shall leave out, for reasons of space, consideration of the disruptions caused in the sugar industry, of the depression in the ship-building yards. There is also something to be said of interference with indigenous systems of medicine and education. I shall end with salt, for it became a symbol of oppression and, later, of rebellion.

Before East India Company rule, the production of salt was free, for trade or private use; it was also necessary: R.J. Forbes has noted how the nature of a vegetarian civilization is always reflected in its profuse use of this almost basic commodity. The Company first imposed a tax on the salt trade; later it made salt a monopoly and increased the price. The corresponding revenue was enormous – already in 1789, it totalled 7 million rupees. In that year, production of it in secret was blessed with penalties. In 1791, informers of clandestine manufacture were awarded one-fourth of the proceeds recovered. In 1812, the total proceeds had reached 11½ million rupees. By 1844, the cost of producing salt was one anna per maund, but the tax on it was two rupees (32 annas). In 1883, W.S. Blunt wrote in his diary:

> The police are empowered to enter houses night or day and, on their accusation of there being a measure of earth salt in it, the owner of the house may be fined fifteen rupees, or imprisoned for a month. If the villagers send their cattle to graze anywhere where there is a natural salt on the ground, the owner is fined or imprisoned, and the salt is thrown in heaps and burned. The cattle are dying for want of it, and the people are suffering seriously. . . . In the Deccan, its pressure is more galling, because natural salt lies on the ground, and the people are starved of it as it were in sight of plenty. In several villages which I passed the ryots told me that they had been reduced to driving their cattle by night to the places where salt is found, that they may lick it by stealth.[46]

Blunt notes that a kind of leprosy had already begun to prevail along the coast, and that the police continued to collect and burn all salt found in its natural state above the ground. In 1883, the salt revenue netted six million sterling for the British. In 1930, the salt revenue netted the same authority £25 million out of the £800 million still being taken out of the country.

In 1930, too, when Mahatma Gandhi wished to begin a new campaign of national civil disobedience, he began with salt: all he had to do was to trek down 200 miles to the Arabian Sea, there stoop and pick up a few grains of salt from the pans, and the entire nation ignited. Later, after a jail term, as he sat for negotiations with the Viceroy, Lord Irwin, he was handed a cup of tea: from a small bag hidden under his shawl, he removed a bit of salt and put it into his cup: "to remind us," he remarked smilingly, "of the famous Boston Tea Party."[47]

CHINA

China, unlike India, was never a full-scale colony and thus, never drawn into world commerce to the same extent as India. But semi-colonialism attempted two different phases of exploitation: one was the attempt to find a market for Western, particularly British goods, which failed. The other, following on the first, was the accumulation of profit through investment, particularly in railways. Ultimately this failed too, when in 1949, China unilaterally cancelled all its debts and refused compensation for investment within its borders by foreign companies.[48]

In the first quarter of the nineteenth century, the position of the European nations in China was roughly what it had been in India before 1748. However, the industrial revolution in Britain had given that country, between 1815 and 1848, a pre-eminent position in the East. If change soon appeared, it did so mainly in 1834, as the East India Company's monopoly of trade was abrogated by Parliament and the field left open to private merchants.

That these merchants were ready to demand "private and untramelled enterprise" need not astonish us, for the opium trade was still labouring under the restrictions of illegality; on the other side, England's industries were crying out already: "Obtain us but a sale for our goods and we will supply any quantity." If the clash was inevitable, the excuse was shocking: opium. After the Opium War (1840-42) and the Treaty of Nanking (1842), the prospect of the sale of British goods to the most populous nation in the world (and nearly in terms of a monopoly) excited the British imagination to the limit: little did the merchants realize that the prospect was nothing more than a chimera. China was not a colony, like India.

For, the value of British exports in China in 1850, inspite of the special privileges of the five Treaty ports, showed no increase over that of 1843. In 1854, on the contrary, they were less. The China firms of the period speak of "depression", of the "unpromising aspect of things", and the "wretched position of your (China) markets". In June 1850, Jardine & Matheson, the most powerful firm in the China trade, reported:

> Our fast monthly advices informed you of the unfavourable turn our markets had taken for imports. This we confirm and advise serious fall in cotton yarn and shirtings, vessels with further goods causing glut that for a long period will not be easily got over.[49]

Perturbed, a select committee of the Commons even considered lowering the duty on tea in the English market so as to increase China's import capacity. More bizarre disappointments included that of the Sheffield firm which sent large quantities of knives and forks to a people who had better reason to use chopsticks, and that of another London firm which despatched a number of pianos in the expectation of "a million Chinese ladies wishing to acquire a Victorian accomplishment".[50]

The China merchants, unable to understand why three hundred million Chinese did not appreciate the quality of Lancashire goods, began to imagine hidden reasons for it: they now began to demand an extension beyond the treaty ports and direct dealings with the provincial authorities to the exclusion of the central government. "Our trade with China", the Manchester Chamber of Commerce declared, "will never be fully developed until the right to sell and purchase is extended beyond the ports to which we are now restricted".

Thus, the Second War (1858-60) and the treaties of Tientsin and Peking, which gave the Westerners twelve new Treaty ports, diplomatic access to Peking, and open navigation on the Yangtze. The results, however, continued to be bitterly disappointing: exports from China were still over 10 million pounds, and *legitimate* imports about 3 million. The balance was met by opium ($4\frac{1}{2}$ million) and bullion. As the consular authority, Alcock, pointed out:

> When the Treaty of Tientsin was made, the cry for more ports and the opening of the Yangtze to Hankow was equally unanimous and loud. Their desire was granted and with what result, let the universal bankruptcy, the nearly total transfer of foreign trade into native hands, and the unoccupied land at all ports. . .attest.[51]

Now the merchants were sure that they must deal, not even with the provincial authorities, but directly with the consumers themselves. As Panikkar puts it:

> Their openly expressed desire was that the whole country should be enlarged into a vast treaty port, with all authority vested in local officials, in dealing with whom the Consuls were to be given the right of calling up the gunboats as a final argument. They pleaded frankly for the establishment of a protectorate at least over the Yangtze Valley and promised in that case that Lancashire would have an unlimited market, and that "all the mills of Lancashire", as Pottinger said, "could not be making stocking stuff sufficient for one of its provinces".[52]

It is not that English consular officials were not aware of the real situation. An Assistant Magistrate at Hong Kong, who made an analysis of English commercial prospects in China, pointed out to the Foreign Office if it did not seem strange that ten years after all restrictions had been removed, China yet did not consume one-half of what Holland did:

When we opened the sea-board provinces of this country to British trade ten years ago, the most preposterous notions were formed as to the demand that was to spring up for our manufactures. Our friends in Manchester and their counterpart on the spot here. . . seem to have all gone mad together upon the idea of an open trade with "three or four hundred millions of human beings".[53]

One consul observed, after an experience of ten years, that "with the exception of our own domestics I have never yet seen a Chinaman wearing a garment of our long cloth, who had to get his daily bread by his daily labour". Other officials warned that "any hope of supplanting the sturdy household thrift of the Chinese" was unwarranted. What these officials indicated was but part of the truth: the Chinese still had a flourishing cotton industry. In fact, in the first quarter of the nineteenth century, raw cotton was by far the biggest legal export to China. In 1827, for example, the East India Company brought to Canton not less than £470,000 worth of raw cotton; and private merchants, £700,000 worth. As Mark Elvin writes, "Between 1785 and 1833, the single province of Kwang-tung imported on average from India each year *six* times as much raw cotton as all Britain used annually at the time of Arkwright's first water-frame."[54]

From the period between 1867 and 1914, cotton manufactures did indeed rise from one-fifth of total imports to more than one-third, yet the major increase was in yarns from India, and later, Japan. These imports might have stunned the spinning industry, but they boosted the weaving section. At any rate, there was also conscious opposition to cotton yarn itself. In the last decade of the nineteenth century, Shantung having become a German "sphere of influence", cotton yarns poured in and caused severe unemployment among local cottage units. The peasants knew perfectly well, according to an American missionary, that "before foreign trade came in to disturb the ancient order of things, there was in ordinary years enough to eat and wear, whereas now there is a scarcity in every direction, with a prospect of worse to come". John Gittings puts this situation down as one of the causes of the rise of secret societies, and the Boxer Rebellion itself.[55]

As with India, however, so with China too, the country's image as a source of manufactures was gradually weakening. Between 1884 and 1914, China's export receipts tripled, but the joint contribution of silk and tea had fallen from four-fifths to one-third of the total trade. The increase in the export receipts was due to the rise of what is generally known as "muck and truck" goods: beans, bristles, eggs, feathers, hides, matting, oil-seeds, strawbraid, and other miscellaneous products of low unit value: these constituted three-fifths of total exports in 1904. As Simkin notes:

> China had, indeed, ceased to be Asia's major source of manufactures and, like the rest of Asia, was reduced to exporting foodstuffs or raw materials, not very important ones at that.[56]

DE-INDUSTRIALIZATION QUESTIONED: INDIA

Thus far, the de-industrialization picture. Its extent, however, has been disputed in recent years – there is a great deal of evidence that indicates that if de-industrialization did take place, its impact was restricted to certain places and regions, particularly trade centres, and since these were highly visible and accessible, any distress affecting them was easily recordable.[57]

Critics also point out that foreign trade formed insignificant aspects of the economies of India and China. They also indicate that a closer study of areas like Gujerat, for example, may show totally different results, too much attention having been given in the past to the events in Bengal.[58]

I prefer, however, to examine the issue a wee bit differently: how did Indians and Chinese in general react to the British impact? My impression is, they reacted quite well, and with a great deal of business acumen. In other words, they adapted to new pressures as readily and as ingeniously as normal human beings do.[59] And where the British had no impact at all, the Indian economy continued to busy itself with the activities it had always busied itself with: after all, India's geographical size should not be underestimated. Sheer distance and inaccessibility, lack of communications probably played a great role in restricting and limiting British influence to certain regions.

As a matter of fact, the early phase of European expansion and participation in Asian trade did raise the productivity of the Indian economy. The entry of the Dutch, British, French, Danish, and Swedish merchants expanded Asian and European markets. In an article on European commercial activity and the organization of India's commerce and industrial production between 1500 and 1750, Professor T. Raychaudhuri arrived at the following conclusions:

To sum up, the impact of European commerce with India on a competitive basis was in many ways beneficient. New markets were opened for Indian exports and the existing ones further deepened. For the limited areas supplying the staples of exports, this meant an increase in production and probably also in productivity, partly through the extension of the putting-out system as well as the localization of industries. Thus, in certain parts of the country at least, the possibility of further significant changes in the volume, technique and organization of production had been opened. But the initiative in innovation remained throughout in the hands of certain foreign companies of monopolistic merchant capital whose interest in reorganizing production was necessarily limited. . . . Certain new techniques in dyeing and silk-winding were introduced by European experts working for the companies. In short, within the limits already defined, new elements of efficiency were introduced in production, probably resulting in an increased productivity.[60]

When exports of cotton goods began to arrive and flood the Indian markets, the Indian weaver seems to have grasped quite eagerly the new opportunity of buying cheaper yarn. Henry Gouger, an Englishman who had set up a factory near Calcutta, with a hundred looms powered by steam, discontinued the working of the looms, and switched the steam-power to the making of yarn: he found the latter more profitable. He himself thus produced, on his own testimony, about 700,000 lbs of yarn for the Indian market.

However, machine-produced yarn still did not mean the total destruction of hand-spun yarn. In 1930, Arno Pearse, a Manchester man, made a study-tour in India to observe its cotton industry. "It is estimated," he wrote,

that there are in India intermittently at work 50,000,000 spinning wheels (charkas) which yield 48 lbs of yarn per spindle per year, and almost 2,000,000 handlooms.[61]

In 1927, cloth woven by handlooms continued to supply 26 per cent of total cloth consumption in the country. Wherever possible, the craftsman used machine-made yarn. Note the following figures, indicating the use by handlooms of machine-spun yarn:

1925-26	283 million lbs
1926-27	324 million lbs
1927-28	323 million lbs

In 1926, for example, cloth woven on handlooms totalled 1,160 million yards, that produced by mills 1,581 million yards, and that imported from England 1,405 million yards. Up to the 1940s, handlooms still continued to produce a third of total output. The total number of handlooms in the country in 1950-51 seem to have been in the region of 3,125,000. Morris D. Morris even feels it probable "that over the century-and-a-half after 1800 the absolute number of handloom weavers increased substantially", and that per capita consumption of cloth also increased.[62]

In the urban areas, the artisan, still required to supply traditional demands, was willing to be receptive to non-traditional materials. In 1918, the Indian Industrial Commission made the following observation:

> The weaver has taken to mill yarn, the dyer to synthetic dyes, the brass and copper smiths to sheet metal, the blacksmith to iron rolled in convenient sections. The tailors invariably employ sewing machines, and town artisans take to improved tools of European or American manufacture.[63]

In the villages, the artisans continued their traditional activities in more isolated contexts. The blacksmith and the carpenter were indispensable for repairs to agricultural implements. The village potter was essential for a variety of functions. The leather-worker was needed for the manufacture and repair of leather containers and buckets, especially in irrigated areas. And the carpenter was primarily responsible for the phenomenal rise of a form of transportation that grew out of British India and will continue to be in use for decades to come: the bullock-cart.

The design of the traditional cart was improved by an Englishman, who replaced the normal solid wheels with a spoked version. The cart as a whole was made generally lighter, could now be drawn by a single bullock, at the most two, instead of the eight or nine required previously. By 1848, there were already 90,000 bullock-carts in the Madras Presidency itself. In the Sholapur Collectorate, if there were 430 carts in 1834, they had increased to 1,907 in 1846, and to 2,643 in 1850. The increase in the number of carts lowered the price of hiring them. In Madras, in 1838, a cart carrying 300 lbs charged 14 annas a day. In 1847, a cart carried 1,000 lbs for the price of 8 annas. It goes without saying that the decrease in prices made possible a further decrease in the prices of food and materials. The bullock-cart industry's phenomenal rise may be taken as an indication not merely of a vast employment potential that had arisen due to the decline of other industries, but also of the readiness of the economy in the rural areas to adapt to fresh demands.

(In 1975, there were over 13 million carts in India, which carried over 60 per cent of the farm produce from the fields to the markets. The total investment on this form of transport was a stupendous 3,000 crores, compared to 4,000 crores invested in the railways and about 1,000 crores invested in road transport.)

Here we rest; the Indian response to Western technology will be taken up in a later chapter. It is time now to turn to China.

CHINA

The political weakness of the Chinese empire that began to be noticeable after 1840 has often been taken to include a corresponding weakness in the Chinese economy itself. I have already observed, however, how English interests, and later European ones in general, were sorely disappointed by the indifferent response of the Chinese economy to the prospect of foreign imported manufactured goods, and how this response might be seen as but a natural result of a still strong, traditional, productive system.

I have already described some of the elements of this system in chapter three; my aim here is to take account of new studies on the nature of this traditional system as it continued to operate till the advent of the Communist government in 1949. These studies unanimously emphasize the continued strength of the traditional economy, while not denying its decline. Perhaps, the term "decline" is inappropriate. The economic situation was not indeed as healthy as it had been in the early decades of the nineteenth century. But there is no evidence to suggest that after 1940 it had been weakened to such an extent that it could not have been able to operate more or less as it had always done and to show signs of regeneration or adaptability to changed conditions.

This is not to deny that the difficulties faced by the traditional system were becoming increasingly serious: the progressive breakdown of civil order and the pressures of a mounting population needed to be faced. Yet, even this should not be taken to mean that China would have veered in the direction of total chaos or collapse. The common opinion that China and its economy *as a whole* was hopelessly degenerating was almost certainly formed in the light of the perception of scholars concerning the situation, after 1920, in the treaty ports and the surrounding areas, where modernization and national integration according to the Western pattern had been attempted and had failed.

There are two cases of overprojection here: first, the situation, admittedly deteriorating in the treaty ports and points of Western influence and control, has been, without cause, projected as existing for the entire Chinese economy. Second, the greater difficulties facing the entire Chinese economy after the 1920s which led to a state of affairs in which the larger mass of the Chinese people had to make do with a diminishing quantum of subsistence, have warped the perceptions of scholars and convinced them, again without cause, that these phenomena and their attendant results had been operating in China since the period of the Unequal Treaties.

These scholars have also included contemporary Chinese historians, for, like Indian nationalist historians, they too have found it necessary to assert in the service of ideology, that Westerners or foreign capitalists "destroyed" Chinese domestic industry after 1840, a view that is belied by the actual state of the economy and the productive system not merely as it existed as late as the 1930s, as Albert Feuerwerker notes, but perhaps as far as 1949.[64]

G. William Skinner's studies (in three parts) concerning the rural marketing system in 1948 indicates, for example, that even in that year not more than 10 per cent of the traditional market networks in the entire country had yielded to modern or Western trading systems; in Szechuan, in fact, he found that none had so altered.[65] As Rhoads Murphey puts it, this traditional network for the exchange of goods and services "was vigorous and flexible enough to adjust as was necessary to changes over time and population, regional political conditions, local disorder, and new commodities, but was rarely and briefly seriously disrupted".[66]

Murphey argues, and with good reason, that the attempt to introduce a foreign industrial system, because it was done through the treaty ports, was in large measure restricted to the treaty ports themselves:

The two separate systems, traditional-rural and modernizing-administrative, touched one another very little, and their interaction was further minimized by the spatial concentration of the modernizing sector in the few scattered islands of the treaty ports or the political struggles centred in Peking and Nanking. In what seems, especially through foreign eyes, to have been a catastrophically disintegrating China, the vast, predominantly rural bulk of the country, including its "little tradition" sector of livelihood and commerce, was only marginally affected.[67]

One of the prime reasons for the misperceived idea of the general state of the Chinese economy itself is probably due to the fact that very few scholars have taken into proper consideration the sheer bulk of China and what that entailed. When Britain set out on the course of the industrial revolution, its population was a mere 2 per cent of the Chinese population of the same period. The difference in size and population simply makes the two economies incomparable in any meaningful sense of the term.

I have already mentioned in an earlier chapter some aspects of the voluminous trade that the Chinese economy embraced or relied upon: certainly, this large network was not built in a day or even in a century; it had evolved as the country, already facing a hundred million in the tenth century, gradually adapted itself to larger and ever increasing demands. Adam Smith, though his information came from secondary sources, came very near in his appraisal of Chinese trade to what the actual state of affairs involved:

> The great extent of the empire of China, the vast multitude of its inhabitants, the variety of climate, and consequently of productions in its different provinces, and the easy communication by means of water carriage between the greater part of them, render the home market of that country of so great extent as to be alone sufficient to support very great manufactures, and to admit of very considerable subdivisions of labour. The home market of China is perhaps in extent not much inferior to the market of all the different countries of Europe put together.[68]

Not that contemporary scholars with a larger knowledge of China differ from this observation. John Fairbank, in assessing the impact of the Westernized treaty ports, noted that "China was too big a country (with its) great reservoir of inland provinces" (many on the scale of separate European states) "to be easily stirred by a marginal sea-frontier contact with foreign ideas". Or, we might add, with foreign products. Those who, like the Abbe Huc, did make journeys through China, felt often that the influence of foreign commerce "is very little felt in this vast Chinese Empire and this immense population of traders. The trade with foreigners might cease suddenly and completely without causing any sensation in the interior provinces."[69]

The Chinese economy responded precisely as the Abbe Huc had foretold. The Chinese did, for example, accept imported machine-spun yarn to some extent when they found it cheaper, as had the Indian weaver, than hand-spun yarn; they also accepted cigarettes and kerosene, both of which had been absent in the traditional economy. The point, however, I wish to make, and where Huc proved to be correct, is that even these goods came increasingly, especially after 1915, from Chinese producers, using Chinese materials.

It is here that the role of the Chinese entrepreneur is significant. The vast traditional productive system was serviced by an equally large number of successful Chinese entrepreneurs and merchants, whom the Westerners eventually failed to displace. The Chinese merchant controlled the domestic trade, but when opportunities did arise to profit in the context of foreign trade, he was able to bring that too under his influence. As Murphey writes:

> After about 1860, and especially after 1920 (the Chinese entrepreneur) took advantage as investor of the new opportunities for profit offered by foreign innovation in steamships, mining, banking, and factory production in the treaty ports. One estimate gives a total of 400 million taels of Chinese capital invested in foreign enterprise in the late 1880s, by which time the Chinese owned about 40 per cent of the stock of Western firms in shipping, cotton spinning, and banking, and held shares in roughly 60 per cent of all foreign firms in China.[70]

Thus, the Westerners were continually thwarted by this control of the traditional Chinese merchant not merely over the domestic, but also the foreign and treaty port trade. In other words, they were attempting to invade a traditional system which was fully able to meet and beat them at their own game of commerce, on home grounds. There was not merely no need but no room for foreign traders to establish a "modern" marketing system along Western lines or with the participation of "modern" merchants. The existing system was fully capable of managing the country's commerce without outside help.

More revealing is the state of Chinese technology and the question whether it proved to be adequate in meeting the primary subsistence demands of Chinese society itself. Here we must set ourselves squarely against Mark Elvin's thesis of a high-level equilibrium trap discussed in an earlier chapter, and the principal features of which include the following ideas: that the traditional Chinese system produced high agricultural yields, adequate manufacture of most essential goods (especially textiles), an efficient exchange linkage by low-cost water transport in the areas of densest population and production – *therefore,* there was a strong resistance to technological change.

The idea of a high-level equilibrium *trap* presupposes the view of technology and technological development that I have set myself against from the very inception of this book: the principle of technological change for the sake of such change itself. Two civilizations, Britain and China for example, are studied practically in terms of a technological race. One undergoes an industrial revolution after which its technology is assessed as "advanced"; the other, China, is compared then to the first, found backward and stagnating, even after it is admitted that the Chinese system as it exists fulfils its function. Writes Murphey:

> But a great deal of the Chinese reluctance or unwillingness to buy foreign goods or to adopt foreign business methods or technology was entirely rational and not culture-bound: traditional Chinese goods and methods were equal or superior and especially so in cost terms.[71]

In 1886, the Commissioner of Customs at Tientsin making his Report on Trade observed that the "customers of the British manufacturer in China are not the bulk of the people but only those who can afford to buy a better looking but less useful article". He also pointed out that the British commodity was significantly more expensive per unit of weight.[72] Yet, Rhoads Murphey goes on to endorse the Elvin trap argument. He continues:

> But China's relative success economically, if only in keeping foreign competition minimal, helped to mask crucial respects in which China was *technologically* backward by comparison with the modern West and also to buttress resistance to technological change.[73]

After this very sentence, Murphey goes on to actually give reasons why China did *not* find it necessary to accept technological change:

Although it became technologically backward by comparison with the post-eighteenth century West, its degree of *pragmatic success* and *self-sufficiency* made it difficult to change [in that case, *why* change]. Western technology was resisted because it was not easily seen as advantageous, not simply because it was foreign. [Murphey then goes on to show that whenever technology, in fact, was seen as advantageous, it was accepted with alacrity. The most obvious exception, the speed and eagerness with which Chinese merchants took advantage of steamship transport, can be seen as a logical extension of the traditional system, which had also been for many centuries evolving its own increasing commercialization, growing long-distance trade and urban concentration (Emphasis added).[74]

In a recent paper, Mark Elvin has found it necessary to qualify his earlier theories by observing that there *were* methods on hand to revolutionize hydraulic technology and that his high-level equilibrium trap argument does not indeed explain why they were not used.[75] Yet, in *The Pattern of the Chinese Past* he set out to explain the disincentives to technological change or improvement precisely on the basis of the trap argument: this shift, reversal, or plain contradiction can be guaranteed to constitute the heavy baggage of all those scholars who assume that China should have produced technological change for the sake of technological change.

The proper historical question to ask is whether there ever has been a civilization devoted to such a monotonous quest. From the Emperor Vespasian, who when presented with a design for a mechanical contrivance that would easily replace the labour of many men, replied that he did not need it for he had his poor to feed, to the contemporary multinational company, technology has always been subordinated to the function of economic, political, or religious goals: I have emphasized time and again that a technological system must be evaluated in terms of the context in which it operates: Western technology, however "advanced", was, in relation to Chinese technology and Chinese tasks, not "advanced", but irrelevant. This I have tried to show already, but a few examples concerning the condition of the handicraft industry might deepen the quality of our opinion.

For example, as late as the 1930s in Hopei, one Chinese scholar's estimates indicate that small-scale handloom weavers still accounted for four-fifths of the total production. Another study suggests that the areas around Shanghai in 1898 were still using hand-spun yarn in important quantities. Twenty-five miles from Shanghai, near Shashih:

> The raw materials were either hand-spun yarn alone or a combination of machine-spun warp and hand-spun weft. . . . The cloth was sold in Szechuan, Yunnan, Kweichow, Kwangsi, Hunan, and elsewhere, either by them [that is, local shops] directly, or through merchants/travellers (hao-k'o) sent out from these areas to lay in stock.[76]

In other words, the system of production was clearly rational and commercially successful. Even as late as 1913, hand-spun yarn continued to be used on a large scale. The gradual shift in some areas to machine-spun yarn, but within the framework of handicraft weaving processes, is further evidence that if traditional methods persisted, they also manifested a remarkable adaptability to changes in cost brought about by technological innovations.

Feuerwerker estimates that in 1933 the output of handicrafts accounted for 67.8 per cent of the industrial share, and that exports of handicraft products increased from Ch $ 104 million in 1873 to Ch $169 million in 1903 to Ch $444 million in 1930.

> In general, on both theoretical and empirical grounds there is reason to believe that domestic demand for handicrafts did not decline in the twentieth century.[77]

The principal reason for the continuance of the handicraft industry was not the "traditionality" of Chinese cultural habits, but the increase in demand brought about by population increase. This large population, mainly rural, continued to use the product of handicraft industry, which, as Feuerwerker notes, "given low wage rates and the high price of capital, could produce traditional coarse goods at a lower unit of cost than modern industry".[78]

All this is not meant to deny that in some areas, such as Kaoyang, Wuhsing, and Ting-hsien, there was some disruption of traditional handicraft industry; there was. And in areas like Chekiang and Wuhsing, handicraft silk reeling and weaving might have grown and flourished between 1870 and 1920, but then both fell prey to technical obsolescence in the face of rayon and Japanese mechanically produced silk. Yet, as Feuerwerker so aptly puts it:

> Anyone who would claim that the Hunan or Szechuan peasant in the 1930s dressed in Naigaiwata cottons, smoked BAT cigarettes, and used Meiji sugar has a big case to prove.[79]

Finally, it is not worthwhile denying that the Chinese would at some stage have had to face the problem of industrialization itself. It is doubtful, however, whether industrialization, if forced on the economy from outside and without Chinese political control over its development would have worked towards the benefit of the large mass of the Chinese people. After 1949, for example, China inherited a modern cotton textile industry which had grown since 1890, and also a traditional hand-weaving industry, which despite the growth of its modern counterpart, had still survived. The Chinese Communists immediately destroyed the basis for the traditional industry by eliminating surplus labour in the rural areas.

Thus, contrary to what happened in India, the modern textile industry did not, between 1949 and 1972, destroy or displace traditional labour by flooding the internal market with cheaper goods against which the latter would soon find it difficult to compete: the problem in China was solved by eliminating the basis of the problem itself. Labour was no longer "wasted" on tasks that could be taken over by mechanized industry in a socialist state, it could be devoted to more constructive tasks. And the reader will bear in mind that in a large number of other industries, unlike in cotton, handicraft industry continued to contribute in rationally perceived ways to the newly vitalized productive system after 1949.

CONCLUSION

I hold it to be a major historical blunder to ask why a particular society did not produce any great desire for economic growth, just as I hold it misleading and wrong to inquire into why a society did not make the transition from traditional to modern science. The historical fallacy is committed when the historian or theoretician subsumes the experience of alien cultures under a theory that he has constructed out of the experience of his own society while it was responding to new problems.

Thus, it is often claimed that the aristocratic elites of most societies, including those of India and China, exploited their lower classes and squandered the surplus on wasteful consumption and luxury goods. Little thus went into productive use or improving technology, so the theory goes. Economic growth therefore was bound to suffer. Thus, we are told that the Mughal state apparatus was parasitic in the truest sense of the term: it was more a regime of warlord predators than an agrarian bureaucracy. How far is all this truth? Dharampal has come up with some startling answers.[80]

Most of the controversy centres round the tax or rent that the peasants had to pay to the political authority. The ancient Indian texts like the *Manusmruti* placed it as one-twelfth (or less) to one-sixth of the gross produce; the practice of the Vijayanagar Rajya is stated by British scholars to have been one-fourth; the 1820s' reporting about the Rajasthan area is also one-third; but some of the mid-eighteenth century accounts regarding Bengal mention its being one-fifth or less in the early eighteenth century. The British determined the land revenue payable at some 50 to 60 per cent of the total produce, that is, more than half.

According to the date of *actual* receipts, however, for the periods immediately preceding British rule, these were at times not even 10 per cent of the computed revenue for particular areas. Dharampal argues that there was therefore a great disparity between what was the theoretical figure demanded as revenue by the political authority and what was actually paid to it.

The major problem, according to Dharampal, and one that has received no attention at all it seems, is where did all this tax or rent go? One would assume that it went to the concerned government and in the case of the areas under Mughal rule to the Mughal treasury. The available published data of receipts regarding the Mughal times put the actual receipts of Jahangir at about 4 per cent of the computed income and that for the reign of Aurangzeb, between 10 and 20 per cent.

The key to this problem seems to lie in the early British reporting on revenue, for Bengal pertaining to the period 1760-90; for Madras, 1750-1800; and for areas in Rajasthan, between 1820-30. The picture that seems to emerge indicates a predetermined division of the sources of public finance among several groups of recipients, the central political authority (i.e., the sovereign authority over the area) being only *one* and that, too, not the most important in terms of the proportion received.

Taking 1,000 as the total gross produce from agriculture and manufactures, in 1750, Dharampal estimates the several allocations as follows:

I. Actual Producers		700
II. Religious, Cultural and Educational Institutions and Individuals		100 of which,
a. Exclusively religious	40	
b. Cultural	40	

c. Educational	20	
III. Economic Services and Police		75 of which,
a. Economic Services	60	
b. Police	15	
IV. Militia and Political Aristocracy		75 of which
a. Militia	60	
b. Aristocracy	15	
V. Central Authority		50
Grand Total		1,000

The proportions under II and III were invariably allocations of definite sources of revenue made *at the primary level itself*, that is, within the village or taluk, perhaps also the district. The greater number of recipients in these categories were individuals, though it may have been that the greater proportion of these receipts were set aside for institutions like the larger temples, choultries, madrassahs, mosques, dargaahs, poligars (heads of area police), etc. The tax-payer paid his tax directly to the allocated recipient. The number of individuals and institutions, for example, that were classified as religious and charitable as late as 1788 was computed to be about 25,000 in what was then the Rangpur Division alone.

With the arrival of the British, the situation was gradually changed: the actual receipts of the central political authority increased from nine- to sixteen-fold of what it had been about 1750. Taking the 1830 gross produce again as 1,000, the division of it may be considered tentatively as follows:

I. Actual Producers	350
II. Religious, Cultural and Educational	15
III. Economic Services and Police	20
IV. Militia and Political Aristocracy	25
V. Central Authority	590
Grand Total	1,000

In such a manner was the fine and intricate relationship between private men and public service undermined: but this is not my point. What I wish to disabuse is the almost universal notion of exploiting elite and exploited peasantry, the former ever ready to squander the fruits of the latter's labour. There was and has always been exploitation in human history. I do not excuse it, this entire book is against it. Neither can I for that matter then excuse historians and economists for seeing Indian and Chinese history purely in such naked terms, when there was much more to it than mere obstacles and incentives to economic growth, the twentieth century's own sacred cow.

CHAPTER SIX

The Renewal of Chinese and Indian Technology and Culture

> Culture has proved to be the very foundation of the liberation movement. Only societies which preserve their culture are able to mobilize and organize themselves and fight foreign domination. Whatever ideological or idealistic forms it takes, culture is essential to the historical process. It has the power to prepare and make fertile those factors that ensure historical continuity.
>
> – The late Amilcar Cabral

The period this chapter covers begins with 1850; the issue is the indigenous Chinese and Indian attempts at technological and cultural independence *vis-a-vis* the West. It would be a truism to say, of course, that the nature of the two nations' historical pasts determined their different approaches to the acceptance and import of Western technology. The depression, however, that both cultures underwent in their evaluations of themselves seems to be similar. It is also instructive to study the Chinese and Indian attitudes to technology and culture after political independence.

It is important to remember, at once, that such elements of Western technology as were imported between 1850 and 1947 into India and 1870 and 1949 into China, were not really important as far as their actual contribution to the economies of the period can be assessed. In the light of the events that took place in industrial production after 1947/49, however, they do assume considerable importance and significance, in so far as they did determine in part the quality of the Indian and Chinese responses as a whole to the question of Western technology after independence.

The figures deny outright any crucial importance to modernized industry in both lands in the colonial/semi-colonial period. According to Feuerwerker, the total share of Chinese and foreign-owned modern industry in 1933 accounted for only 2.2. per cent of net domestic product.[1] (Murphey estimates a 1933 figure of 3.4 per cent of net domestic output.[2]) Both figures propose little significance, especially if we note further that in 1933 the output of factories, handicrafts, mining, and utilities in China constituted just about 10.5 per cent of the net domestic product. If the population of China estimated in 1933 at 500 million is correct or credible, then the *working* population in agriculture was 212.30 million, and the *working* population in non-agricultural occupations, 46.91 million. Of the latter figure, 12.13 million were employed in handicraft industry, 1.33 million in factories, 0.77 million in mining, and 0.04 million in utilities.[3] Thus, not merely did industry, including handicraft industry, occupy a small role in the Chinese economy, but within the industrial sector itself modern factory production occupied a very small place. Feuerwerker concludes:

> Although twentieth-century China experienced some industrial growth in the treaty ports, some development of mining and railroad transportation, the extremely small numbers engaged in those occupations even in 1933 suggests that the occupational distribution of China's population as a whole had changed very little from what it had been at the end of the Ch'ing dynasty.[4]

The Indian picture at independence is not very much different. In 1947, large-scale factory production in India employed less than 3 million people as compared with 12½ million in small-scale industry and handicrafts in a total labour force of 160 million.[5] In brief, in either case, both China and India were not in any sense close to being industrial powers.

The semi-colonial feature of Chinese society and the total subjection of India *vis-a-vis* Western dominance was bound to have an equally significant impact on the elements of Western technology that were first felt necessary in these economies. Take the Chinese case.

CHINA AND WESTERN TECHNOLOGY

Twenty years after the disastrous Opium Wars, Chinese minds finally found themselves ready to enter a new phase, the "Restoration" of the T'ung-chih period. The first scholar to devote his attention to this attempt at "resuscitation", Feng Kuei-fen, immediately proposed that what the country speedily needed was a calculated scheme to adopt Western military technology:

What we then have to learn from the barbarians is only one thing, solid ships and effective guns. When Wei Yuan discussed the control of the barbarians, he said that we should use barbarians to attack barbarians, and use barbarians to negotiate with barbarians. . . . In my opinion, if we cannot make ourselves strong (tzu-ch'iang) but merely presume on cunning and deceit, it will be just enough to incur failure. Only one sentence of Wei Yuan is correct: "Learn the strong techniques of the barbarians in order to control them."

Feng went on to suggest the establishment of a shipyard and an arsenal in each trading port; the invitation of "barbarians" to teach bright Chinese students and artisans military technology. He devalued the importance of the civil service examinations, proposing that one-half of the scholars should henceforth apply themselves tc the pursuit of manufacturing weapons and instruments and imitating foreign crafts. He showed himself well aware of Japan's use of foreign military hardware to remain strong. Most striking, finally, is his insight about how best the Chinese might go about the proposed task, one which we shall argue is an essential condition for any new nation to develop its own technological programme today:

Some suggest purchasing ships and hiring foreign people, but the answer is that this is quite impossible. If we can manufacture, can repair, and can use them, then they are our weapons. If we cannot manufacture, nor repair, nor use them, then they are still the weapons of others. When these weapons are in the hands of others and are used for grain transportation. they one day can make us starve. . . . Eventally, we must consider manufacturing, repairing, and using weapons by ourselves. . . . Only thus will we be able to pacify the empire; only thus can we play a leading role on the globe; and only thus shall we restore our original strength, and redeem ourselves from former humiliation.[6]

Most of the documents in the earlier section of Fairbank's and Teng's *China's Response to the West* betray this selective approach in the Chinese desire for Western technology. For example, the chief architect of the Ch'ing dynasty's victory over the Taiping Rebels, Tseng Kuo-fan, in his writings shows a gradual progress from the Confucian preoccupation of writing eight-legged essays to a more intense interest in Western ships and guns.

As early as 1853, he prepared a memorial about the need for a naval force to improve China's defences. His interest extended from a keenness to study Western methods of training soldiers to getting his people to imitate and manufacture foreign guns he had purchased. In 1855, he set up a small arsenal in Kiangsi; in 1861, he moved to Anking and set up another arsenal there and a shipyard. About the theory of self-strengthening that occupied the Chinese intellectuals of his time, he wrote:

If we wish to find a method of self-strengthening, we must begin by considering the reform of government service and the securing of men of ability as urgent tasks, and then regard learning to make explosive shells and steamships and other instruments as the work of first importance. If only we could possess all their superior techniques, then we would have the means to return their favours when they are obedient, and we would also have the means to avenge our grievances when they are disloyal.[7]

Tseng Kuo-fan continued his work building arsenals: he established the Kiangnan Arsenal himself and continued to urge the building of steamships. Equally important in this scheme of defence-building was Li Hung-chang, one of the most powerful officials to emerge in the lower Yangtze provinces in the 1860s: his first contribution was to take practical steps to secure Western arms. Li's position of leadership in China was in *Yang-wu* or foreign matters, which involved not merely diplomatic relations but the borrowing of Western technology. Fairbank and Teng summarize his wide range of activities and influence:

The foreign-style enterprises were begun mainly for military purposes and followed one another in a logical sequence. To suppress the rebellions and for coastal defence, there were first, the establishment of arsenals and shipyards, and the building of forts and vessels. Secondly, technicians were needed to make these weapons and so schools were established and students and officers sent to study abroad. Since modern defence required modern communication and transport, the construction of telegraph lines and the organization of a steamship line were undertaken. Eventually, since modern defence also required money and material resources, a cotton textile factory was established, and coal, iron and gold mines were opened.[8]

Thus the problems of industrialization were sought to be solved through a progression of ideas beginning with the strategy of "using the barbarians to control the barbarians" to employing Western arms, to the conviction that Western arms must be manufactured in China, produced by Chinese themselves, and that Chinese must be instructed to make them; further, it was necessary that the Chinese take seriously a training in Western sciences in general. Both new institutions and elements of an important infrastructure were essential for the progressively enlarged aims. For example, in 1872, the China Merchants Steam Navigation Company was founded to compete with British shipping in China: it was soon to need a Chinese coal supply independent of foreign imports. In 1878 the Kaiping coal mine, a forerunner of the Kailan Mining Administration north of Tientsin was opened to meet this need. And the earliest surviving railway in China was later built to connect with the Kaiping mine.

The emphasis on defence needs and the technological complex it entailed stands out more clearly, perhaps, if we observe that the modernization of textile production, an industry with less obvious strategic value, moved more slowly in comparison. Tso Tsung-t'ang put up a woollen mill in Lanchow, Kansu in 1878, with the aid of the Germans, but it did not flourish after his death in 1885. And an 1882 planned cotton mill at Shanghai prepared by Li Hung-chang was not set up till 1891: the new cotton mill, the first in China, burned down in 1893 after only a year of successful operation. Li soon re-organized it, this time on a larger basis. At Wuchang, Chang Chih tung set up a cotton mill in 1891; a cement factory was attached to the Kaiping coal mine, a match factory and flour mill were also opened.

Heavy industry got started even more slowly. Chang Chih-tung opened the Ta-yeh iron mine and the Hanyang Iron Works in Hupei in 1890, with German technical expertise, but these projects never developed into the big industrial complexes that Chang had hoped for. A list of the major developments during the entire period between 1863 and 1891 betrays an overwhelming interest of the Chinese in the strengthening of the country through defence means.[9]

INDIA

In India, on the other hand, such a preoccupation with defence (which affected the choice of Western technology accepted not merely in China but also in Japan) was conspicuous by its absence. But then India was already a colony of the British, unlike China and Japan, both of which continued to maintain a political identity *vis-a-vis* forces from "outside".

Isolated attempts were made by the English to start industrial projects in India. Thus, the first steam engine was imported from England already by 1820 through the Christian missionaries of

Serampore near Calcutta, for their paper mill.[10] I have already mentioned Henry Gouger and his steam-powered spinning mill. Under the patronage of the Madras government, the Porto Novo Iron Works were started by Mr. Heath in 1825: by 1840, these works employed 43,000 workers, but they were finally closed down by the government in 1861.

True to colonial policy, the British government had little interest in a serious industrialization programme as such, and it is useless and unwarranted at this stage to criticize it for any neglect in this regard. Recommendations of the Famine Enquiry Commission, 1880, to encourage industry and sponsor technical training officially were ignored for forty years.

Early this century, the Government of Madras tried to break this pattern and set up a Provincial Department of Industry. Its activities are said to have "roused the opposition of the local European commercial community, who interpreted them as a serious menace to private enterprise and an unwarrantable intervention on the part of the State in matters beyond the sphere of Government. . . ."[11] Consequently, Lord Morley, Secretary of State for India, disapproved of the entire project, disbanded the Department, the leather tannery which had provoked the largest outcry was sold, and the experimental handloom shops abandoned. Similar attempts on the part of the colonial government elsewhere met the same discouraging attitudes. And yet, some form of industrialization was inevitable. When it did begin, it followed a pattern very different from the one Feng Kuei-fen had envisaged for the Chinese.

Sardar K.M. Panikkar was one of the few Asian historians courageous enough to observe that the British entry into Indian political life in 1757 had been made possible through a group of rich and influential Indian traders who had discovered the lucrativeness of British trade for their own purposes. The eighteenth century, in fact, had seen the consolidation of powerful Indian banking houses which handled revenue remittances and advances for the Mughal Empire itself, and also the Nawab of Bengal, the East India Company, other foreign companies and Indian traders, and which also carried out arbitrage between Indian currencies in different areas and of differing vintages. These indigenous banking houses were largely pushed out of function by the British.

The first textile mill in India was started in 1851 in Bombay by Indian capitalists who had made their wealth through trading with the

British and had acquired some knowledge of English. This mill and the others that followed it were launched with some essential financial and managerial help from the British trading companies: a situation precisely the reverse of what had happened in 1757. By 1945, indeed these indigenous capitalists and their modernized textile mills had absolutely displaced British textile imports.

The rise of Indian capitalism and foreign industry went therefore, unlike in China, hand-in-hand. The goal of the Indian capitalist was to cut into the profits of British goods sold in India by producing these goods in the country itself through means of production imported directly from an industrialized Britain: by 1877, several cotton producing areas like Nagpur, Ahmedabad, and Sholapur had their own textile mills: most of them concentrated on the production of coarse yarns which were sold domestically or exported to China and Japan. India's textile industry preceded Japan's by twenty years and China's by another forty. Consequently, there was now a gradual change in the nature of its exports.

In the 1850s, India's exports, instead of the finished products of industry, consisted merely of raw materials such as jute, wheat, cotton, oil-seeds, and tea. At the same time, its market remained flooded with British goods, including luxury items such as silks and woollens, leather and leather products, cabinetware and furniture clocks and watches, earthenware and porcelain, glass and glassware, paper, pasteboard, stationery, toys, requisites for games, scents, cigarettes, carts and carriages, and, later, bicycles and motor-cars.

But after 1879, as Indian capitalist production established a stronger hold within the economy, things began to look up once again. Thus, the proportion of manufactured exports to total exports from India rose from 8 per cent in 1879 to 16 per cent in 1892, and to 22 per cent in 1908; the proportion of manufactured imports to total imports also fell, from 65 per cent in 1879 to 57 per cent in 1892, and to 53 per cent in 1907.

Two important events gave the industry a shot in the arm. From 1905 onwards, the *swadeshi* movement, a national boycott of British goods in favour of Indian ones, aided not merely Indian textile firms and other industries, but also Indian insurance and banking. The First World War reduced imports from Britain a few years later. Indian finance reached out to control foreign managing companies suddenly short of funds. By the 1920s, for example, majority ownership of the largest organized industry, jute, had passed into Indian hands. Indian firms also tightened their hold on the home market for textiles and steel.

In 1914, Indian cotton mills produced only one-fourth of the mill-made cloth consumed in India; by 1935, their contribution had risen to three-fourths. By 1946, India had even begun to re-export textiles, as I noted earlier, to Britain itself. The point to keep in mind is that the inherent conflict between the traditional and the modernized textile industries, unlike as happened in China, remained unsolved, even though control had changed hands.

THE RENEWAL OF CULTURE

A nation that sees its technological ability as low in relation to another is almost certain to devalue its own image of itself. Historically, this is evident in the cultural histories of both China and India in the period of Western dominance. If the technological power of the Western nations offered their members the easy and gratuitous assumption of the cultural power of Western civilization itself, both India and China were forced to undergo periods of intense cultural depression, when they began to entertain serious doubts concerning the viability of significant elements in their cultural traditions.

Fairbank and Teng identify in fact two main study areas concerning the China of 1839-1923: one of these, the socio-economic, I have already discussed. The other, they term "psycho-ideological", and it is precisely this problem that concerns us now. Fairbank and Teng attempt to place the issue in the following terms:

> [The psycho-ideological approach] is concerned first of all with the traditional Chinese ideologies – the systems of values and beliefs which supported and sustained the old order. Secondly, it is concerned with the slow and many-faceted breakdown of those ideologies under the corrosive influence of Western power and Western ideas. Thirdly, it seeks to analyze the absorption and adaptation of those Western ideas which interacted with persisting elements of the old order. In short, this approach studies the ways in which modern Chinese have sought to create new systems of value and belief to replace the no longer adequate ideology of the disintegrating traditional order.[12]

Fairbank and Teng conclude that the experience of modern China must be studied through psychology as well as economies and social organization. More important, they realize that this entire problem is in a very real sense the problem of a minority: the selections in their work "represent the Chinese elite, not the common people". Or, "as the inherited institutions and habits of thought lose validity, intellectuals experience tension and anxiety, greater hopes and fears". Alternatively, it is equally obvious that the majority of the Chinese people (and of the Indian too) did not suffer this strong period of self-doubt: for them, tradition still continued to provide a sure guide for action, and traditional technics a possibility of meeting their primary needs.

That the elites of both these large nations underwent a crippling sense of inferiority and humiliation *vis-a-vis* Western culture is so obvious that it need not be presented in great detail. What is necessary, however, is to distinguish the Chinese context from the others, for the Chinese feeling of inferiority *was* different from that of others.

For unlike India, China had a vigorously self-conscious cultural nationalism, national identity, and a long tradition of an integrated state and culture for more than two thousand years before the arrival of the Westerners. The latter's point of entry, however, was restricted to the mechanism of the treaty ports, and served in fact merely to reform and sharpen traditional Chinese insistence and belief of their self-sufficiency and self-satisfied identity and to provoke a distinct response to Western models.

In contradistinction to this, colonial rule in Asia might be said to have given rise to nationalism and a national consciousness almost *de novo*. Most politically conscious Asians, lacking their own national tradition, and already beginning to note serious inadequacies in their local traditions, were ready to accept a Western model and to see even a clear path to progress under British or foreign rule.

Thus, India. like Ceylon and Japan and even Southeast Asia, experienced a long identity crisis. In each country, to differing degrees, indigenous attitudes, cultural styles, techniques, patterns of thought, notions of political and social organization – the whole stuff of traditional society – were found wanting and to varying degrees rejected, directly or indirectly, in favour of outright Western models or a modified hybrid.

China never had such an identity problem during the semi-colonial period: the foreigners provided no model for progress, instead an example of banditry and true barbarianism. Thus, the felt dominance of the West created a firmer commitment of the Chinese to their own cultural tradition, and heightened the sense of China's identity rather than destroying it: every document of the period emphasized the *Chinese* sense of crisis, not the cosmopolitanism of China. Second thoughts about Confucianism, the discovery of science and democracy, even the abandonment of cultural membership values, did not involve any abandonment of cultural membership or real loss of identity. No one in China wondered, in the turmoil brought about through the Western impact, in the learning of English, in the rejection of blocks of the Chinese past, who he was, as nearly all other Asian nationals did. China was in danger, writes Rhoads Murphey, "but not Chineseness".

> China suffered infinitely more, but at least in part because it never even briefly flirted with the idea of not being Chinese; this was perhaps a losing game after 1850, but to play any other was nevertheless unthinkable.[13]

Thus, at least up to 1949, Western technology provided an ideological function: in revealing to the Chinese their backwardness, it cut a deep wound in the Chinese psyche: China was "humbled". The imperial centre was politically, technologically, and ideologically *powerless;* the increasing regionalization of Chinese politics was blamed by the nationalists on the treaty ports, and those Chinese who "collaborated" with the foreigners in the treaty ports, especially those who adopted foreign techniques and attitudes, were understood as traitors. Significantly, the revolution would come from Peking, never a treaty port itself: the first political activities to come out from that "spiritually" clean centre took the form of protests and boycotts *against* the treaty ports and all they represented, "national humiliation". It was this new group, divorced both from the traditional order and from the treaty ports, which inherited modern China.

INDIA

The abject feeling of inferiority in India was the result of a different set of circumstances, brought about principally by total subjection to British rule. Unlike the Chinese, Indians adapted at first to the roles that Empire required. The psychological and moral effects of British conquests and Indian subjection gradually spread and deepened. The disappearance of the warrior element in Indian society (the Kshatriyas) marked the disappearance too of basic components such as courage and encouraged more superficial doubts among Indians about their technical ability to do anything about the overthrow of British rule.

British rule succeeded in making clear to the Indians themselves that they lacked power, and it strengthened the imperial opinion that qualities of passivity, weakness, and cowardice were in fact norms of Indian culture *and* character. The process was no doubt aided when the British concentrated on providing educational and related service opportunities that required the tamer skills and temperament of the office rather than the scepter and sword. On the other hand, Britons were led to think that the superiority of English power and culture was an inherent rather than a historical phenomenon. What is even more surprising, the devaluation of Indian culture led to a contempt for the Indian *physique.*

> The physical organization of the Bengali is feeble even to effeminacy. He lives in a constant vapour bath. His pursuits are sedentary, his limbs delicate, his movements languid. During many ages he has been trampled upon by men of bolder and more hardy breeds. . . . His mind bears a singular analogy to his body. It is weak even to helplessness for purposes of manly resistance; but its suppleness and tact move the children of sterner climates to admiration not unmingled with contempt.[14]

This is a passage from John Strachey's *India,* written at the turn of this century and a standard training assignment text at the time for Englishmen undergoing probation in the Indian Civil Service. As the Rudolphs note, what is most significant about these distinctions is that most nationalist Indians half-accepted them, and no ideology legitimizing superiority-inferiority relations is worth its salt unless it wins at least a grudging assent in the minds of those dominated.

> Within twenty years of the deliberate exclusion of United Province Brahmans from the Bengal Army because of their leading role in the rebellion of 1857, the idea that Brahmans lacked fighting qualities had become prevailing opinion. Reading recent history back into an undifferentiated past, Indians came to believe that they lacked valour and moral worth. . . . Why inferiority in arms, technology and organization, circumstances related to particular historical contexts that may be reversed, has led colonial peoples to more essential conclusions about themselves is not entirely clear. The fact that they frequently did come to such conclusions was one of the most degrading consequences of colonialism.[15]

There were two clear responses aimed at meeting this excruciating problem and I take them up each in turn. The first is associated with Mahatma Gandhi, of course, and he largely solved the issue. The young Nehru, often sceptical of Gandhi's political strategy and tactics, conceded again and again his effect on the nationalist regeneration:

Much that he said we only partially accepted or sometimes did not accept at all. But all this was secondary. The essence of his teaching was fearlessness and truth and action allied to these. . . . So, suddenly as it were, that black pall of fear was lifted from the people's shoulders, not wholly, of course, but to an amazing degree. . . . It was a psychological change, almost as if an expert in psychoanalytic method had probed deep into the patient's past, found out the origins of his complexes, exposed them to his view, and thus rid him of that burden.[16]

That description is indeed apt, for Gandhi did probe the nation's historical sub-conscious, with his unique sensibility both for the nightmare terrors of the Indian psyche and for its commonplace daytime self-doubts: the shape he gave to the national movement for independence, above all the technique of *satyagraha,* had much more than strategic significance; it provided a path for action that "solved" some problems of Indian self-esteem arising from acceptance of the negative judgements of Englishmen.

It is easy to misunderstand the nature of this "truth-force" or *satyagraha,* especially if one is conditioned to accept the Western definition of courage as stressing masterly aggressiveness or heroic acts of self-assertion. The error in the mis-perception of the nature of non-violence has usually been to see it as a failure of will or a surrender to fatalism. Gandhi, in fact, turned the moral tables on the British definition by suggesting that aggression was the path to mastery of those without self-control, non-violent resistance the path of those with self-control: it is best to turn to history at this stage to provide a striking example of the difference.

In the previous chapter, we mentioned the salt march: the "Long March" in 1930, 200 miles to the Arabian Sea to collect salt against British prohibition: Gandhi had staked everything on this Salt Satyagraha, and it would be in this campaign that the quality and essence of his non-violent methods would appear. There were about 2,500 volunteers that finally entered the Dharasana Salt Works From here, our description passes over to Webb Miller, a British journalist, whose account of what happened has passed into the realm of the classic:

In complete silence the Gandhi men drew up and halted a hundred yards from the stockade. A picked column advanced from the crowd, waded the ditches, and approached the barbed-wire stockade. . . . Suddenly at a word of command, scores of native policemen rushed upon the advancing marchers and rained belows on their heads with their steel-shod *lathis* (sticks). Not one of the marchers even raised an arm to fend off the blows. They went down like ninepins. From where I stood I heard the sickening whack of the clubs on unprotected skulls. The waiting crowd of marchers groaned and sucked in their breath in sympathetic pain at every blow. Those struck down fell sprawling, unconscious or writhing with fractured skulls or broken shoulders. . . . The survivors, without breaking ranks, silently and doggedly marched on until struck down.

They marched steadily, with heads up, without the encouragement of music or cheering or any possibility that they might escape serious injury or death. The politice rushed out and methodically and mechanically beat down the second column. There was no fight, no struggle; the marchers simply walked forward till struck down.[17]

The very absence of violence incited the police to viciousness: feeling defenceless in all their superior equipment, all they could think of doing was what seems to "come naturally" to uniformed men in similar situations. They bashed in the volunteers' skulls and kicked and stabbed them in the testicles. "Hour after hour stretcher-bearers carried back a stream of inert, bleeding bodies."

Two months later, the Indian poet, Rabindranath Tagore, wrote triumphantly to the *Manchester Guardian* that Europe had now lost its moral prestige in Asia. But he had missed the point, as Erik Erikson points out: praising the Mahatma, Tagore had written that weak Asia "could now look down on Europe where before she had looked up". Gandhi, notes Erikson, might have said it differently: Asia could now look Europe in the eye – not more, not less, not up to or down on. And, adds the psychoanalyst, "where man can and will do that, there, sooner or later, will be mutual recognition".[18]

Side by side with the Gandhian response was another, more tame and compromising; in fact, it came from a new class of Indians, many of whom rejected Gandhian ideas and by the time of independence had constituted themselves as the new leaders, and men of control. Every society has its own version of an elite, and it is instructive to spend some time analyzing the "algorithm" of the Indian one: it is representative for most of the Southern countries. But first, a little detour.

The classical *theory* of the Indian caste system is the *varna* system, which comprises the rather well-known categories – Brahman (priest),

Kshatriya (warrior), Vaishya (trader, merchant), and Shudra (labourer). What actually exist in Indian society, however, are not these four strict divisions, but a number of castes and sub-castes, and this is the *jati* system: the two, theory and practice, should not be confused.

Yet, there is a line that divides: members, for example, who have been initiated with the sacred thread ceremony, are "twice-born" and belong to the three of the higher castes. The mass of sub-castes that form the Shudras do not possess this privilege.

Put this way, the caste system would seem to lack and restrict social mobility. In actual practice, considerable mobility is possible: normally, the lowest castes adopt some of the key features of the upper castes and pass themselves off as "twice-born". M.N. Srinivas, India's leading sociologist, termed the phenomenon "Sanskritization":

> The tendency of the lower castes to imitate the higher has been a powerful factor in the spread of Sanskritic ritual and customs, and in the achievement of a certain amount of cultural uniformity not only throughout the caste scale, but over the entire length and breadth of India.[19]

To be more specific, the lower castes normally adopt certain of the rituals of the higher castes, like the wearing of the sacred thread and at the same time try to command the services of Brahman priests and use Vedic rituals. A Munda tribal, for example, who had established himself as a local raja, attracted Brahmans to his court, who Sanskritized his rituals and manufactured a Rajput genealogy for him, thus legitimizing his position in the higher ranks of the caste scale. Such channels of social mobility are not restricted to India, but whatever flexibility the caste system did have seems to have been apparently lost when the British administration bureaucratized and set the categories.

The social mobility afforded by the caste system is reminiscent of the democratization of gentlemanly culture and standing in eighteenth-century England that preceded the shift from a society of relatively closed ranks and orders to one of relatively open classes. Daniel Defoe, as well as other pamphleteers of class and manner, by providing the English with popular literary instruction in the art of becoming and being a gentleman, and by celebrating that status, facilitated the expansion of the ideal to those previously excluded from it. The only difference is such a democratization process came from above, while caste mobility is a movement from below.[20]

High up in the caste hierarchy, however, a similar movement was taking place, as the leading sections of the twice-born castes underwent "Westernization", a process in which they began to use English and adopt the occupations and cultural style of the West. They were continuing the process, though, in a different framework: this time, the English had become the dominant caste, and just as the leading sections of India had once become "Persianized" under the Moghuls, they now found it beneficial to change their dress, language, and diet, not in order to demonstrate that they had become modern so much as to show they were emulating the culture of the dominant class.

Contrary to the Dutch in Indonesia, the Portuguese in Goa, and the Spanish in the Philippines, the English established themselves as a separate ruling *caste*: like other Indian castes, they did not inter-marry or eat with the lower (native) castes. Their children were shipped off to public schools in England, while they themselves kept to their clubs and bungalows in special suburbs known as cantonments and civil lines. The close contact with the caste system strengthened British snobbery: the British civil service, with its tradition of generalists and Brahmanical status of the administrative class, is practically derived from the Indian model.

The new Indian elite was the product of English-created opportunities, and these opportunities were more easily exploited if its members successfully embraced Western, or more particularly, English ideas and manners. Political power was sought to be added to traditional priestly, commercial, and literary power.

Every one of these new elites would be found in the port cities of Asia: by mid-nineteenth century, it was obvious (with the aid of hindsight, of course) that all of them had already become fundamentally Westernized, and that it would be these Westernized groups emerging in Calcutta, Bombay, Madras, Karachi, and Colombo who would inherit and shape South Asia's development after independence.

By the end of the century, the same pattern was apparent in each Southeast Asian country (except Laos and Cambodia), where the new national elites were again to be found exclusively concentrated in the colonial ports. The point is not so much that these elites were so alienated from the large majority of their countrymen, but that in inheriting the power to shape the modernization of their countries, they would simultaneously inherit the power to choose the kind of technology with which they would seek to accomplish their purpose. And thereby hangs a tale.

The tale concerns the kind of technology normally attractive to elites of this sort. A concrete example of the process of acceptance of kinds of technology in an actual village situation in India will make it quite clear.

In the previous chapter, we presented a summary of Mahmood Mamdani's study of the village of Manupur, where we discussed principally the indirect influence of the moneylending class on the state of agricultural technology. We observed, with Mamdani, that any kind of technology that might prove useful to the general, agricultural section of the population in the village was not yet acceptable as the farmer or tenant realized that the possible benefits of its use would not accrue to him, the tiller, but to the moneylender, who was often also a member of the Brahman aristocracy.

It should therefore come as no great surprise that the first radical departures in technology in Manupur in the first half of this century were those that benefited the Brahmans. Mamdani makes a careful distinction:

> The innovations introduced into Manupur in the 1920s and 1930s were the radio, the handpump, the watch and the bicycle, and their ownership was confined to the three leading brahman families. Significantly, these technological changes were on the side of *consumption* not *production;* were external to the agricultural sector; and were, in their context, luxury items.[21] (Emphasis added.)

Changes in agricultural or productive technology were practically non-existent; iron gradually substituted wood in cartwheels and plough, and the older leather-bucket system gave way to the Persian wheel. The latter change affected only a minority of peasant farms, and all of their owners had not had to mortgage their land to the moneylender.

> As long as the peasant producer remained subservient to the money-lender, and the money-lender reaped the rewards of increased productivity, the system remained highly resistant to technological innovation. For technological change to take place on the side of production, it was necessary that the producing classes be free from the grip of the parasitic money-lenders. A dynamic economy was not possible without a dynamic dominant class.[22]

Change was possible only after an Intensive Agricultural Development Programme (IADP) in 1960 provided loans to farmers for repaying their debts and for making changes in agriculture, and in the process undermined the material basis of the Brahman aristocracy. The effect on productive technological changes was soon discernible after that. By 1970, 75 per cent of the farmers had taken loans to purchase their tubewells. Mechanical chaff-cutters replaced hand-cutters for making fodder for cattle.

By 1970, there was not one farmer who did not use chemical fertilizers: manure was used merely as a supplement. In 1964, electricity enabled electric motors to run tubewells and chaff-cutters. The watch proliferated among the agricultural class, as transactions in the nearby town of Khanna were determined there by the fact that the shops opened and closed at specific times. New seeds were rapidly accepted. By 1970, 75 per cent of the families owned at least one bicycle and 24 per cent a radio: both necessary, one for travel to town, the other for listening in to broadcasts from government about agricultural matters. Most significant, products of technology that made the least impact were those that could be called semi-luxury or luxury items, all entailing wasteful expenditure.

Electric irons in six households and coal-irons in thirty others were necessary as the newly prosperous farmer and trader would not be respected in town on transaction visits unless they emulated the ways of the town; the clerk and the school teacher were expected to wear ironed clothes to work. The least-adopted technological product was the ceiling fan: a luxury item, owned by 8 per cent of the households, the most prosperous farmers, testifying to the phenomenon of a new rural bourgeoisie in the making. As for the Brahmans, their material base having disappeared, their religious dominance was also destroyed. As an elderly Brahman complained:

> (Once) only Brahmans performed priestly functions. The Jats (rural labouring class) couldn't even begin to move the plough until the Brahman had performed certain rurals. Then, we were held in esteem and needed. Now, Jats say they are not Hindus and don't believe in our rituals.[23]

Let me now draw a very large conclusion, in a wider context. Only 5 per cent of the population of India today earns more than 350 rupees. The top three per cent controls one-fourth of the country's expenditure. A third of all the money circulated in the country remains within a restricted circle comprising at the most ten per cent of the population. This influential elite is still fundamentally Western in its tastes: it is also the only realistic market for consumer goods. The non-elite, 90 per cent of the population, lacks any substantial purchasing power. Indian public and private industry, plus the multinationals, are both one, in so far as profit remains their principal motive, in catering to this elite; the kind of technology invested in or transferred to India is thus vitally distorted in its nature and purpose. In India, as is the case in practically every Southern country in the world (there *are* a few exceptions) one finds a major portion of industry heavily prejudiced in favour of the production of consumer goods: luxury items and non-essential goods, which are actually non-productive in the context of the real needs of their populations. In sum, these elites behave in exactly the same manner as far as technology is concerned as the Brahmans did in Manupur in the 1920s-1930s.

The dependence of tastes on the part of these elites on the West – still looked up to – is repeated elsewhere, particularly in the sphere of ideas. Indian scientists, technologists, even professors of philosophy are deeply involved in issues that concern their counterparts in America or Europe, in order to gain a place in the "international community". An Indian novelist can make the grade in India only after he has published his novel first in the West. Knowledge may be power, but in the context of the dependence relationship that exists between the Southern nations and the West, power itself has become knowledge: Western scholars have, seemingly, not merely established *the* criteria for Popperian objective knowledge, but the knowledge that they produce is more objective than most other efforts. Part of this book has therefore concerned itself with examining the ethnocentrism implied in the production of a great deal of Western "objective" knowledge. Jacob Bronowski was merely an excuse.

An excuse to establish the emergence of a new phenomenon: an international caste system, the amplification of the caste system from its limited moorings in India into the larger face of the world. If the Brahmans once prescribed "dharma" (caste duty, the freezing of occupation roles) to the castes below them, this was to their own advantage and since they had the superior role, they even re-wrote the scriptures to enforce it. If they closed their eyes to lower groups attempting to "sanskritize" themselves, this is because they again realized that the process itself reinforced and legitimized the further acceptance of the system's authority. What might indeed have frightened them would have been a situation in which the lower castes had suddenly lost all interest in "sanskritization", thereby signalling a rejection of the worthiness of the Brahmanic model and its authority.

In almost similar terms, the industrial world continues to prescribe to the world's Southern nations how best they might help themselves. Walt Rostow's proposal was one such attempt; today, a new (but equally specious) proposal is making the round: interdependence. Specious, for underneath the strategy remains the same, to convince the Southern nations that their welfare is safe so long as they remain intimately tied to the Western capitalist framework. Part of this framework are a small minority of elites in the Southern nations who by "sanskritizing-westernizing" themselves have in the process, accepted the continuing legitimacy of the Western paradigm's dominance. China's breakthrough brought about a period of fear that was, as I pointed out, even reflected in the work of American and European scholars.

GANDHI AND MAO TSE-TUNG

Mahatma Gandhi distinguished himself from the rising Indian elite by his absolute rejection of the Western paradigm of human development.[24] By 1947, however, he had been displaced from any position of effective political influence, for, as Nehru himself admitted, "much of what he said we only partially accepted or sometimes did not accept at all". Nehru and his colleagues particularly refused to entertain Gandhi's suggestion that the Congress party be dismantled to form a rural reconstruction movement.

Gandhi's disillusionment with the leaders of the Congress party was so total, he went to the extreme of prophesying that the near future would find him leading a civil disobedience movement against *them*. How far the new Indian elite had distanced themselves from his thinking is evident from the following extract from a letter of the first President of India, Rajendra Prasad, justifying his living in one of the most expensive mansions the world has ever known: Rastrapathi Bhavan. Prasad wrote:

> People belonging to the group which had adjusted itself during Gandhiji's time find it difficult to understand why it should be necessary for anyone to have anything in excess of what he had during Gandhiji's time, and they

honestly feel that there is a decline from the ideal if there is any change from the austerity standard set by Gandhiji. All that can be said about such critics is that standards set during a time of struggle and accepted by the people cannot be expected to be equally acceptable when the struggle is over, and they are wrong in expecting such standards to be maintained at all times and in all circumstances.[25]

For Prasad, the struggle was indeed already over; for Gandhi, it was only beginning. The Gandhian conviction was, however, cut short by an assassin's bullet, though the core of Gandhi's thought evolved into the Sarvodaya Movement, a basically rural reconstruction force and the mention of which brings us to the examination of a very important issue.

Political scientists have often compared Mahatma Gandhi to Mao Tse-tung: as personalities, their role in their respective periods of influence has, indeed, been similar as both men succeeded in mobilizing vast masses of men and women in the direction of a single purpose. There are differences, however, adequate enough to place the two men into two vastly different categories. Gandhi's rural reconstruction movement, which he intended to put into operation after 1947, outside the sphere of politics and government, has little to do with Maoist principles of revolution and social change. Instead, the Sarvodaya Movement may only be realistically compared with the Rural Reconstruction Movement symbolized in China by the Confucian holy man, Liang Shu-ming, in the thirties.

Liang was the founder of an influential project at Tsoup'ing in Shantung, which proposed local self-government (ti-fang tzu-chih) or rural self-government (hsiang-ts'un tzu-chih), which implied political organization below the hsien level and thus mass participation in government.[26] The project was in keeping with the atmosphere during the period of the Nanking regime, when reformers decried rural bankruptcy and collapse and argued and worked for rural renewal.

Liang's movement, which took root with the founding of the Shantung Rural Reconstruction Institute in 1931, is to be distinguished from two other Chinese developments of the same period. It was basically a civilian movement, different thus from the Kuomintang modernization programme with the help of a bureaucracy, that proposed change and improvement from the top downwards. The latter kind of development is characteristic of the rural development official programmes of the Indian government after 1947.

Secondly, Liang's "old" school at Tsou p'ing was to be distinguished from the "new" school of the Yale-educated Jimmy Yen, the latter a philanthropic institution, funded with foreign aid and ideas. Yen worshipped Western culture, and was convinced that the 5,000 years of Chinese history and culture were the real "enemy" of the Chinese development programme. He wanted to create modern, "scientific" rural villages, and in his ideas was not very different from many development experts from the Western world working in the Southern nations today.

In contradistinction to both, Liang refused any outside aid and ideas, and argued for a strategy where only the peasants could save themselves. His was therefore a typical Chinese-style rural reconstruction ideal, whose fundamental spirit involved the use of the power of character formation to solve practical problems. The Shantung Institute was the main intellectual and spiritual centre of the rural renewal movement, just as Wardha was, and continues to be, the equivalent centre of the Gandhi movement.

Like Gandhi, Liang was interested in a grandiose goal in the Chinese context – a new Chinese culture and society with implications for the whole world, which would aim at reaping the benefits of modernization while avoiding the evils of the West, but within the framework of Chinese sage wisdom and organization. Like Gandhi, Liang further argued that children should have a "basic" or useful education: knowledge of basic literacy, agriculture, general science, hygiene, and civics. The training schools he established involved Gandhian habits of continuous moral scrutiny of members and fellows, and even morning meditation. Like Gandhi, Liang proposed cooperative societies and "trustee" ownership.

Liang's movement collapsed with the Japanese invasion in 1938. Before that, however, it had already been heavily criticized, by both the "Yale" school and the Communists. The proponents of "wholesale Westernization" observed that Liang was an unscientific, old-fashioned restorationist, who had not quite understood the modern world, a charge Arthur Koestler once made about Gandhi. But it is the criticism of the Communists that enables us to see the differences in Liang's approach and Mao's and therefore, between Mao's and Gandhi's.

The Communists termed Liang at once the most reactionary and the most progressive of the leaders interested at the time in agrarian socialism; for he had realized that both imperialism and the Chinese warlords were the true causes of Chinese peasant poverty. At the same time, his approach did little to eradicate these peasant enemies. They noted that his cooperatives received loans from city banks, where foreign interests were heavily involved, and that his plan for the resuscitation of rural industries was collapsing precisely because of the continued imports of foreign goods. As long as imperialism remained, his plans for cooperative socialism were in essence fanciful; and as long as his movement did not prevent the consolidation of the rule of landlords and big peasants (in any case, it was not directed against them), it was bound to be revolutionary in a conservative way, for peasant interests are not identical with those of their more privileged neighbours, and *vice versa*. Willem Wertheim has made a similar point about Gandhi's tactics.[27]

Yet, there were also similarities between the Maoists, Liang, and Gandhi which we cannot underestimate, but they serve to uphold the radical nature of the differences. More important, they concern the nature of the renewal of culture that has occupied us in some portions of this chapter.

In the first place, Liang shared with Mao a peculiarly Confucian faith (as Gandhi in the Indian context) in the influence of the human environment and the efficacy of intimate group contact in rural and intellectual improvement. All three, in fact, conceived of internal virtue (be it a rectified heart, a proletarian consciousness, or the Indian ideal of Brahmacharya) as linked with external political, military, and economic success. According to all three, the good society was to be achieved by continuous spiritual transformation of the whole society, a never ending moral drama (for example, Gandhi's newspaper autobiography), which would solve their country's economic and political backwardness and, at the same time, avoid the dehumanization of urban bourgeois society: all three were apprehensive that selfish mundane desires might extinguish the spirit of sacrifice. Mao and Liang distinguished man from animal and Chinese man from Western man precisely on Chinese man's renewed ability to act against his material self-interest for moral reasons. Mao differed from Liang and Gandhi in that for him the purpose of self-sacrifice was the nation-state and its people, whereas, for Gandhi and Liang, sacrifice was an end in itself, not a means. It was an expression of the *tao,* and only secondarily a means for the benefit of the collective.

It was the Maoists in the end who realized Liang's ultimate goal: the revival and reintegration of China based upon an impassioned man's commitment to a common ethic, "a religion that was not a religion", as Liang often described Confucianism. The Maoist revival also incorporated many other aspects of the Liang programme: the emphasis on small rural industry, local self-reliance, independence from foreign ideas and aid, small-group dynamics, and agricultural development through the peasants themselves.

Probably, the most striking difference between the Maoists and the Liang-Gandhi movements lies in the fact that the former effected a radical re-distribution of power and property before 1949. One has only to read Jan Myrdal's *Chinese Village* report to see that the bulk of the landlords must have been liquidated during the course of the Long March itself. Both Gandhi and Liang had not yet reached this stage, and not having reached it, would soon come to discover that most of their efforts to produce social change would ultimately be blocked. The Chinese people, on the other hand, at least the majority of them freed from an exploitative past, were now ready to work with the new revolution, for they now had a stake, a vital stake, in its continuance.

POST-INDEPENDENCE RESPONSES: INDIA

The attitudes of the Indian and Chinese governments, constituted after 1947 and 1949 respectively, to the role of Western technology in the context of their individual development programmes, will be examined in this last section of the present chapter.

There is a great deal of truth, briefly, in the notion that Chinese xenophobia and Indian xenophilia perhaps explain the major substance of these attitudes. How much of Chinese xenophobia was a matter of choice or necessity after the break-off of Soviet technical and financial aid in 1960 is difficult to decide. But to Indian xenophilia first.

In India, it was already evident after the 1850s that Indian capitalists would import sophisticated machinery regardless of the consequences this would have on the further disintegration of the village industry. The argument used to support the import said that a mechanized textile industry, for example, would keep profits within the country itself. The First World War reduced British imports, and did bring large profits to the Indian capitalists: in 1920, 35 companies which controlled 42 mills declared dividends of 40 per cent and more; ten of these companies controlling 14 mills paid out a dividend of 100 per cent; the dividend of two mills shot up by 200 per cent.

The Second World War raised profits to greater heights: 61 Bombay mills, whose paid-up capital totalled 139.3 million rupees, made a net profit of $6\frac{1}{2}$ times this amount in five years. The Indian textile industry even then employed half a million workers, as against 10 million still engaged in the handloom industry. The mills needed 104 persons to produce a million yards, while the traditional industry employed 6,250 persons to produce a similar quantity.[28]

After 1947, British control was replaced by Indian capitalist control, or step-in-your-shoes nationalists, as Angus Maddison terms them. The history of the Indian government's response to industrial development is a history of continuous compromise with the Indian private capitalist sector. But both, India's private *and* public sectors (as Ward Morehouse has indicated), have in turn mortgaged the country's long-term technical capacity build-up by relying too heavily on easily available foreign companies and governments.

The Indian capitalist originally was not a solitary entrepreneur, but formed part of a tightly-knit, family group, reinforced by caste and communities. In 1951, for example, the Birla family controlled 245 companies and was substantially interested in another eleven. Of the 195 public companies in the complex (eight of which were then among the fifty largest in the country at the end of the decade), seventy-one were engaged in investment financing, thirty-four in trade, twelve in cotton, eleven in engineering, nine each in sugar and tea, six in property, five each in jute, publishing, and managing agencies, four each in food and insurance, three in plastics and in glass, two each in coal-mining, power, non-ferrous metals, and transport, and one each in mining other than coal, rayon, chemicals, paper, construction, fireclay, and banking. Another of the largest groups, the Tatas, had 68 public companies that ranged over at least twenty industries in 1951; the Dalmia-Sahu-Jain group's 63 public companies over eighteen industries: the Bangurs' 33 over fifteen industries; the Thapars' 30 over twelve industries, and so on.[29] Today, some of these companies have already become multinational companies, operating quite successfully in other areas of the Southern world.

Most of these companies, producing principally consumer goods, were weakened in their output capacities by a certain amount of technological backwardness; two foreign exchange crises brought the foreign multinationals into the picture, since it was thought that more foreign investment would increase the country's exchange position. This it did not. And in fact, it was the first foreign exchange crisis of

1957 that forced the Indian government to take seriously its decision to industrialize. This is not to be wondered at, for the opinion of some influential economists such as Asoka Mehta had been that India could best serve an ancillary role to Japan.

Because the industrialization policy came on suddenly, as it were, and suffered improper planning, the modern sector soon fell prey to an unhealthy "technological mix". J. Eddison scrutinized the paper industry as early as 1952. He wrote:

> Most foreign-made equipment was planned and constructed for use under contrasting climatic conditions, and with dissimilar raw materials, differing grades of chemicals, and more highly trained workers than are to be found in India. In consequence, this equipment often gives unexpected difficulties and generally operates at lower efficiency than it would in its native land.[30]

Since the machines were not produced domestically, the industry contained a jumble of machine types: twenty-seven of the forty-six were British, seven German, four Belgian, and the rest Swedish, American, Canadian and Japanese. Nineteen were produced by one manufacturer: no more than five of the remaining came from any one firm. There was, therefore, no proper service organization in the country, and the paper producers were often forced to fabricate, as far as they could do so, their own substitute for worn or damaged fittings. This could be said of most of the other industries. For example, it is often claimed that India produces a greater variety of internal combustion engines than any other country in the world, a situation that can only be considered a built-in check to more standardized production and better service in the long run.[31]

Much ado has been made of the pressures the government has brought on Indian industrialists to find import substitutes. The question is whether the substitues in any way escape the problems carried in by the originals.

The fact that this foreign technology, original or substituted, is irrelevant to the larger productive needs of the Indian economy is realized or admitted with difficulty. In fact, the transfer of Western technology itself is in large part an irrelevant problem, for apart from cotton textiles, India is not a mass-consumption economy; only ten per cent of the population enjoy the bulk of industrial production. The industrial nations, on the other hand, are interested in the transfer of technology for it continues to be profitable and promises to be essential in the long term. The Indian capitalist collaborates, because even though foreign technology may be a drag on the country's resources, high profit margins within the country's restricted market make the enterprise lucrative. The xenophilia pervades the entire scene: from the government through the industrialists, down to the urban consumers, particularly the affluent sections.

From the beginning, in fact (1947), the government realized vaguely that it would have to work towards some form of industrialization: the Indian Plans were consciously patterned on the Soviet industrialization model. For this, the government found it had very often to purchase technology or turn-key projects from abroad, either from the market economies or from the socialist ones; in the early stages, most collaborations were sought for the electrical and machine goods sectors. Later, however, as momentum lagged, government found it needed technical aid in mining, petroleum, machine tools, and the production of all kinds of metals and chemicals. In preparing its contracts for the purchase of industrial goods, it often gave preference to firms with foreign collaboration.[32]

An example will suffice: the fertilizer industry. From its very inception, this industry has proved to be an example of how foreign business interests, with assistance from Indian collaborators, have succeeded in sabotaging authentic indigenous development of industrial capacity. This has happened despite the existence of a competent band of Indian scientists and engineers in the Planning and Development division of the Fertilizer Corporation of India, who could well have carried this particular industry to self-reliance.

In fact, official quarters did entrust responsibility for the development of the industry to these technicians in the late sixties and early seventies. But for some reason not at all known to reasonable men, the local technicians were soon spurned, and affairs climaxed with the decision to return to turn-key arrangements with Japanese interests for the construction of new fertilizer plants, even though the Indian capability in the field was thereby left underemployed, and despite the readiness of international agencies to give credits for the construction of the fertilizer industry in the public sector.[33]

The new government headed by the members of the Janata Party is no different in its attitude: already Indian technicians are being ignored. The most surprising development is the recent move to call in the Intermediate Technology Development Group based in London to advise government on the setting up of small-scale cement plants, even though the country has a fine cement research institute.

Does India need such technical aid? In 1961, there were 28,000 degree holders and technical personnel unemployed; by 1971, their number had arisen to 2,88,487. This number, large as it is, does not include lesser educated labour, much less unskilled and uneducated labour. There is further little scientific and technical knowledge and experience that India needs that is not any more available in the country itself. The psychological damage of technical dependence is worse: the attitude that "someone else is better qualified and can do it for me" saps a people's energy and initiative, reduces the role of local people to bystanders. As in the industrial revolution once, the poor and unemployed have become a burden in most of the new nations. In India, their number has not changed, but increased absolutely since independence. The tragedy is this need never have been so.

To argue, therefore, as Professor Jan Tinbergen has done, in the recent RIO report to the Club of Rome that multinationals have no other alternative but to engage now in research and development that will have direct relevance to the problems of poverty in the Southern nations is to make a profound error, for in doing so, he merely endorses the reality of continued technical dependence, no longer defensible, and further implies that the people in the Southern nations cannot solve their technical problems by themselves.

CHINA

Unlike as it has been made out in books, China did not, like India, sit down like Rodin's thinker to discover the best policy it had to follow. But perhaps its choice concerning its own technological development was rendered easier precisely because before it even got down to facing the question of technology, it had provided for itself a firm context for any action in that direction through basically non-technical means: the effecting of a truly revolutionary redistribution of power, through which it acquired the capacity to choose and implement its policies effectively on behalf of the peasants and workers – whereas India did not. This indeed is a brave statement on the part of an Indian, but it should be understood as constituting merely the beginning of an analysis, not the end.

The key indicator of the validity of such a proposition is to examine the manner in which each society (China and India) provides for its lower income groups, whether unemployed or underemployed. John

Gurley's statement below appears to be amply justified by the evidence on China available today:

The basic, overriding economic fact about China is that for 20 years she has fed, clothed, and housed everyone, has kept them healthy, and has educated most. Millions have *not* starved; sidewalks and streets have *not* been covered with multitudes of sleeping, begging, hungry and illiterate human beings; millions are *not* disease-ridden. To find such deplorable conditions, one does not look to China these days but rather to India, Pakistan, and almost anywhere else in the underdeveloped world. . . . The Chinese – all of them – now have what is in effect an insurance policy against pestilence, famine and other disasters.[34]

John Gurley might have added that rich Indians have indeed grown richer, and that no Indian equivalent of the Chinese "insurance policy" is available to 90 per cent of the Indian population.

No useful purpose is therefore served by beginning a study of modern China by dubbing it a totalitarian state. John Fairbank is a fair example, when he writes of the Chinese people having "succumbed to a totalitarian communist faith". The same Fairbank once asked his readers to forget the horrors of the Opium War and the Western nations' total involvement in it:

The adjustment of modern China to the multi-state system, her proper functioning as part of the world community, will remain incomplete until this sense of grievance at her modern history is exorcised by a rational perspective on it.[35]

This, indeed, is a little difficult to accept. But Fairbank, I noted, is one example. Lewis Mumford and Simon Leys cannot mention China today without speaking of totalitarianism at the same time.[36] Sinologists who see China "from the other side of the river" are rare and far between: Charles Bettelheim, Joan Robinson, Joseph Needham, John Gittings. The last is one of few to accept that the Chinese have the right to determine their own view of human development:

At the same time China *is* a different society with different ideals and social goals from those of the capitalist countries of the West, and no useful purpose is served by trying to pretend that the Chinese are "just like us", even if they are similar in many respects. Intense political struggle, sometimes leading to violence, has been an important part of the mechanism which has driven the Chinese revolution forward from the early years of land reform to the Cultural Revolution. And while everyday life for the Chinese has much in common with our own, the forms of social organization and (much more important) the collective spirit behind them is altogether another matter.[37]

It is generally accepted that the Chinese have attained some of their egalitarian goals, but some Western economists are not quite sure whether these have been achieved at the cost of economic growth. The Chinese have proved themselves willing to subordinate some strictly economic goals to other considerations, but they also rehearse their economics. Other Sinologists are still prescribing to the Chinese what is good for them. Rhoads Murphey, for example, tries to explain away the Chinese decision to de-urbanize industry as a result of the humiliation associated with the foreign experience at the treaty port cities:

> This and the other cardinal sins of "bureaucratism" and "status quo-ism" are no doubt inevitable products of urbanism, but without urbanism the industrialization which Chinese Communism so determinedly wants is *literally unattainable*. Chinese improvements on Western models – for which there is certainly plenty of room – should be sought in other ways.[38]

Alvin Gouldner, in his brilliant article on Marxism and Maoism, has *his* own explanation concerning Mao's lack of enthusiasm for industrialized urban centres. "In one part," he writes, "Maoism's hedging about urbanization is an effort to arrange China's social order so that if it must, it can survive even nuclear warfare."[39] Mark Elvin wonders if the Chinese experiment will not develop deep fissures as the country's advancing industrialization will place it into closer contact with a "corrupting" West. I could go on.

On the other hand, J. Gray has argued that behind the political thought of Mao Tse-tung is a strong economic policy, and that, in fact, the two should not be seen as contradictory.[40] Gray's analysis is in keeping with one of Mao's key talks on Chinese industrialization: *On the Ten Great Relationships*.[41] Charles Bettelheim, in his recent work, has a splendid analysis about how the Chinese have succeeded in improving productivity and in arriving at solutions transcending a narrow technical outlook.[42]

Sinophiles, however, rarely emphasize the trial-and-error approach that has underlied Chinese industrialization strategies. During the fifties, for example, China relied heavily on Russian technical assistance to develop its modern industry. In fact, the Chinese policy of self-reliance might have appeared much later in the day if the Soviets had not suddenly pulled out in 1960.[43] In the early sixties, China's production priorities had changed, calling now for a major expansion of petroleum and chemical fertilizer production: the unfavourable foreign exchange position at the time meant that most of the plant and equipment would have to be produced by the domestic machine-building industry. China did meet the challenge, while India, in a similar foreign exchange crisis, simply sought the easier way out by inviting foreign aid.

The Soviet pull-out probably convinced Chinese leaders that technological self-sufficiency must be promoted at all costs and the role of foreigners drastically limited: this has led to a very considerable indigenous technological base. In India, on the other hand, whether in the public or private sector, it would be difficult to find more than a handful (in the very literal sense) of large factories built with indigenous know-how, not to speak of capital equipment.

In China, for example, the Taiching petroleum complex which began in 1960 soon after Soviet supplies were halted, was soon under way to making the country independent of foreign supplies altogether. Bettelheim writes:

> The result has been that China now holds the world record in terms of international drilling norms. Annual production of crude oil continues to increase by about 30 per cent. In terms of its oil requirements, China is now self-sufficient. Taiching represents for Chinese industry what Tachai represents for agriculture. It points to the socialist road of industrialization.[44]

In India, again, Western-based oil companies got away lightly with Indian socialism or the government's ability to quickly compromise. An opportunity to develop self-reliance in the petroleum industry and reduce foreign permanent dependence was not grasped and enforced as forcefully as, for example, Cuba did.[45]

"Walking on two feet", intermediate technology, the encouragement to innovate and experiment, are part of the broad Chinese consensus favouring a high degree of national and regional self-reliance in the manufacture of machinery and other producer goods. I doubt the Chinese think in terms of technological "autarchy" *vis-a-vis* the industrial nations or even of self-sufficiency. They are interested more perhaps in autonomy or self-reliance, which is a different thing and surely more realistic.

The results of such a policy especially at the local level are bound to be mixed. Thus, reports of thriving factories of Kwangtung's prosperous Chungshan country alternate with reports of failure, as in Hopei's Kaoch'eng country, where efforts to combat drought foundered because locally produced engines turned out to be useless. The Chinese, however, seem to tolerate such failures in their continued desire for more widespread industrial experience as a long-term policy.

Richman, an authority on the industrial experience of both India and China, draws the conclusion:

> The point is that the Chinese can produce practically anything they wish to, though in limited numbers and at great costs in many cases. I feel that Red China has a significant lead – perhaps five to ten years – over India in overall product development and know-how in spite of the considerable amount of foreign collaboration and assistance in India's industrial sector. In general, Red China appears to be substantially more self-sufficient in technology and product development and much less dependent on foreign assistance or imports than India. These are critical factors to be considered in assessing future technological and product development prospects in the two countries, and in predicting their industrial and economic growth potential.[46]

CHAPTER SEVEN

The Logic of Appropriate Technology.

> The idea that there may be alternative technologies in itself implies the idea of technological pluralism in place of the until now almost universally accepted technological monism. In this case each social system and each political ideology, indeed each culture would be free to develop its own particular line. Why should there not be a specifically Indian technology alongside Indian art and why should the African temperament express itself only in music or sculpture and not in the equipment which Africans choose because it suits them better? Why should Russian factories follow Anglo-Saxon patterns? Might there not be an unmistakably Japanese technology, just as there are typically Japanese buildings and clothes?
>
> – Robert Jungk: 1973

Let me begin with Chinese archaeology.

Non-Chinese have been exposed to the view that the Great Proletarian Cultural Revolution of the late sixties was a devastating experience for Chinese archaeology. The Red Guards did, indeed, shout of smashing every relic of tradition associated with the old Emperors and their concubines, and everything tainted also by the West. Was it indeed the virtual end of Chinese archaeology as most Western scholars were ready to believe? Evidently not. For the Chinese had once again seized the opportunity, in turning over their past in the field, to reverse simultaneously the interpretation of that past. As William Watson wrote:

> The anger about the imperial past had spilled over into threatening demonstrations against some imperial sites and treasures, but it had steadied and been chanelled into a more constructive attitude: that the treasures of the past demonstrated the ageless skill and genius of the working class who made them, not the genius of the emperors who had enjoyed them.[1]

In a very real sense, China's past belonged to its people, to *the* people, and it was the people of China, the workers, peasants, soldiers, and archaeologists, who would take a concrete interest in preserving it. The Yukang Cave temples might have spread the message of religious superstition which helped bolster the feudal regimes However, as great works of art and sculpture, they occupy an important position in Chinese contemporary history. They reflect the superb talents of the labouring people of ancient China and remain priceless relics for critical study and assimilation.

Simon Leys does paint a different picture of the *actual* Chinese attitude to their ancient memories in granite.[2] But I am fascinated in the scene that opens up wide with that simple, and sudden, reversal, with the idea of that reversal itself: what indeed happens when we view history from the bottom of the pyramid? The same object now betrays another gleaming consciousness. We have a new fulcrum to move the world, to dislodge the older, rutted, but deeply entrenched view of the world, of the place of technology, culture, and man in the world.

A careful reader of what has been laid down thus far in this volume has probably noticed by now three underlying levels of this 'interpretation from below". At the first level, dealing with technological revolutions of the past, I have argued from a perspective that could, I feel, be identified with the majority segment of a society undergoing those revolutions. A technological revolution, I pointed out, always brought more suffering and labour to the large majority at the base that was forced to participate in and make it possible.

At the second level, I took another elite-non-elite combination; the small, influential, powerful minority that constitutes the Western world was inspected through the eyes of a non-Western. A "sociology of knowledge" about dominant and dependent cultures betrayed non-objective interpretations the dominant cultures held and propagated. At level three, finally, I argued the existence of a third, this time the largest, elite-non-elite relationship, that which holds between the half that lives and "the other half that dies". I spoke of the survival technicians that belong to the latter half, but did not say too much about the half itself.

This final chapter will take each of these levels, now made explicit, and piece together the different elements that they comprise, but which have them, thus far appeared in disconnected places. I will spend very little time with the first level: there is plenty of work being done on it today, and there exists too a corpus of documentation on its various aspects. Neither is the second level the most important one, but it occupies paradoxically a great deal of space. I believe that the issues bunched together here will see an inevitable solution in the coming immediate future: that is the reason why it is interesting, even intriguing, to spend time and space unravelling the threads of that inevitability. The third level is our most important area of concern.

THE FIRST LEVEL: ELITE VERSUS NON-ELITE HISTORIOGRAPHY

I observed earlier that the writing of agricultural history has never been accomplished by those who actually made it all possible, and I mentioned Lynn White's complaint that "not only histories but documents in general were produced by social groups which took the peasant and his labours for granted". One of the few scholars to have made a detailed study of the problem was Esther Boserup, who, unable to make a secure analysis of past events from the literary evidence, reconstructed them instead from the experience that contemporary farmers and peasants go through. On this material did she focus her sharp economic tools to come up with the remarkable results I presented and discussed in chapter four.

The industrial revolution, on the other hand, does not seem to have suffered such step-motherly treatment. E.J. Hobsbawm has much to say about those who doubt the suffering the age produced, which white-collar historians like Landes seem unable to reconstruct.[3] "When trade was normal," wrote a liberal economist in 1840, "about a third of the population lives in terrible poverty and on the verge of starvation. A second third, perhaps even more, earns little more than the ordinary rural worker. Only one-third receives wages which allow them a fairly reasonable standard of living and a little comfort."[4]

I cannot forget to mention Karl Marx and Engels and a host of others that have laboured on the question: I may leave it in safe hands. Besides Marxism, no reasonable theory has come forth to grasp the description of events, and Marx, even Marxists admit, is a bit dated. Wilkinson's work is admirable, and he goes further than Marx, which means, of course, his determinism is more extreme. I agree with *Poverty and Progress* a great deal, for population growth, "the weight of numbers", as Braudel called it, seems to have been one variable beyond our control. There is evidence of social control of population in a great many societies, particularly the smaller cultures. But the larger societies have remained bewildered in the face of any increase in hands and mouths, as Thomas McKeown has shown.

In such a situation, men are bound to have to suffer. Not all men, of course, but most men. Emphasizing all this is the chief merit of Wilkinson's fine model of ecological development. It could be used with profit in other situations. Geertz's analysis of Java and Japan comes very near to it.[5] And North and Douglas provide evidence for earlier periods.[6]

There *is* a theory that comes close to what I would like to emphasize. I am thinking right now of Wertheim's emancipation principle.[7] A few words on the background and content.

In 1960, four good anthropologists got together to present a slim little volume, *Evolution and Culture,*[8] in which they tried to distinguish "specific" evolution – the adaptation of a culture to its natural environment, from a more general evolutionary course for mankind:

> The more specialized and adapted a form in a given evolutionary stage, the smaller is its potential for passing to the next stage... specific evolutionary progress is inversely related to general evolutionary potential.[9]

This is another way of saying that there is an advantage in being backward: a society, for example, cluttered up with a massive infrastructure built for oil energy might find it more difficult to switch to solar energy than a society that is very near being a blank slate, a wisdom Mao Tse-tung recognized. Wertheim indicated the similarities between this theory and the ideas of the Dutch historian, Jan Romein, who talked in terms of the "dialectics of progress" or the "law of the retarding lead". Wertheim criticized both theories with regard to their predictive power and their precision, finally noting that their precision was "hardly greater than that of the biblical saying: And the last shall be the first."[10]

Dissatisfied, he has seen fit to provide his own version of the human story, under the rubric of the "rising waves of emancipation", a hypothesis he has defended with considerable skill and passion. He writes:

> It appears to me that the basic principle underlying the concept of evolution could be understood as a general trend towards emancipation. At the same time, this general trend cannot be separated from an increasing human capacity to cooperate....
>
> The general trend of human evolution, therefore, amounts to an increasing emancipation from the forces of nature....

Emancipation from human domination, therefore, goes hand-in-hand with emancipation from the forces of nature....[11]

The trouble is the emancipation of human domination has not gone hand-in-hand with emancipation from the forces of nature. If it has, this is probably temporary, and in such unusual circumstances that it would be dangerous to generalize. Wertheim formulated his ideas before the Club of Rome report, the oil crisis, and the threat of a permanent exhaustion of the raw materials on which the industrial nations have built their economies and their futures. (He admits that the emancipation of some portions of the globe is intimately tied to the repression of people in other, larger portions of the globe.) But take Robert Heilbroner, one of the leading apologists for capitalism, pronouncing on the future of industrial societies themselves:

In place of the long-established encouragement of industrial production must come its careful restriction and long-term dimunition within society.... Rationalize as we will, stretch the figures as favourably as honesty will permit, we cannot reconcile the requirements for a lengthy continuation of the present rate of industrialization of the globe with the capacity of existing resources or the fragile biosphere to permit or to tolerate the effects of that industrialization.[12]

Heilbroner went further and held the threat of more totalitarian governmental systems that he saw inevitable in the democratic societies of the West, including America. He wrote:

In bluntest terms, the question is whether the Hobbesian struggle that is likely to arise in such a straitjacketed economic society would not impose intolerable strains on the representative democratic political apparatus that has been historically associated with capitalist societies.[13]

Let me try to answer the problem by stressing an aspect of it which we have hitherto ignored – the extent of the institutional changes needed to attain a condition of ecological equilibrium. Central among these changes will assuredly be the extension of public control far beyond anything yet experienced in the West, socialist *or* capitalist. To bring environmental stability, the authority of government must necessarily be expanded to include family size, consumption habits, and of course the volume and composition of industrial and agricultural output.[14]

It is true that Heilbroner's dark forebodings have been decried in some quarters, especially by those who continue to believe that there is no real resource threat in the long run. But he is at least right in calling attention to the fact that whatever be the condition of our resources, neither capitalism nor socialism offer anything like a credible long-term solution to economies dependent on them. The belief that either socialism or capitalism can offer any *serious* solutions to the industrial crisis, to take the crisis seriously, can only be explained as Jason Epstein put it, "as an act of faith, a pious regression to formulas and assumptions that belong to an age that celebrated industrial progress as an unmixed and inevitable blessing and whose ideology was confirmed by the New World's seemingly endless resources. These resources could, one believed, be transformed by human will to produce whatever version of utopia one chose; for socialists eventual equality through social ownership, and for capitalists security for those who adapted best to the conditions of survival."[15]

As I pointed out in chapter one, such a situation must of itself contain the germs of war, for it will always imply a continuous search across more and more frontiers for ever dwindling resources. This is the reason, for example, why a democracy like the United States of America must continue to have those 92 military contracts with an equal number of nations to protect and police its resource routes. This is the reason why, finally, America has produced that height of human backwardness, the neutron bomb, which destroys people, but keeps buildings (and presumably, dollars) intact.

All this is not to underestimate Wertheim's "rising waves of emancipation". I do not think that in regard to the Southern nations he ever adopted that trickle-down reform attitude so characteristic of most of his colleagues in the Western academic world. On the contrary, his writings on revolution, his partisan attitude in favour of the oppressed, his early sympathetic attitude to the People's Republic of China, his unilateral condemnation of the aid social scientists have offered to oppressors, his ability not to mince matters when dealing with the nearly universal, corrupt regimes of the Southern nations, have earned him a sizable following among concerned Southern country scholars.[16] It is a pity that, except for *Evolution and Revolution,* most of his work remains within Dutch borders. I shall return to him when we move on to level three.

THE SECOND LEVEL: THE RELATIVIZATION OF THE WESTERN PARADIGM

The selling of the West has been very similar to the selling of American Presidents: in each case, provided one is not too mesmerized, it is possible to see the Emperor's clothes for what they are. Two views of the West, however, can be gleaned from the literature. One sees the West plagued with as many problems as the Southern nations (perhaps more). Arjun Makhijani has beautifully summarized this view thus:

It ignores the fact that most people in the industrialized countries are trapped in meaningless, dull jobs over which they have little or no control; that they must befoul their air and water by endless consumption on pain of losing their jobs; that most are locked into a transportation system where a good portion of their work time goes to purchasing and maintaining a car so that they can get to and from work; that beyond a certain point increases in the Gross National Product have an inverse correlation with well-being and freedom as witnessed by increasing expenditures on drugs (over and under the counter), by rising crime and increasing police forces and armaments, by the alienation of people in urban areas, and by the way people have come to derive their spiritual satisfaction from the things rather than from people around them.[17]

The other view, of course, sees the West as "advanced" or "developed", and it is this view that has been more *successfully* sold to the people of the Southern nations, or rather to their leaders. Throughout this book, I tried to make an indirect distinction between an actual history of the West and its recent advances to power, and the presentation of the Western paradigm by Western intellectuals and scholars. What are the key elements of this paradigm? As proposed in the most general terms, the West sees its development as having required its transformation from a state of rural, agricultural underdevelopment to one in which the urban, industrial sector dominates. "As urban industrialization accelerates, the relative share of agriculture in the total labour force and in the national product decreases. Higher wage rates motivate migration from rural to industrial-urban areas which, together with capital investment in agriculture, increases the productivity of labour in agriculture to levels approaching those obtained in industry."[18] This pattern of sectoral change has been similar to all nations currently considered "developed": a continuous, absolute, and relative growth of sectors unrelated to agriculture with a concomitant population transfer from rural to urban areas.

Yet, this is merely the basic structure and there are a host of other elements needed to complete the overall picture. In the first place, this sectoral change is seen more or less as a matter of conscious design. As Rajni Kothari observed, the seeds of progress are seen as being sown with the rise of modern science and rationalism in Western Europe, and its intellectual flowering in eighteenth-century "enlightenment". "Everything that followed was an unfolding of the enlightenment – the commercial and industrial revolutions, the growth of representative institutions, rational bureaucracy, the egalitarian ideology, the socialist state, the modern city and its culture."[19] Max Weber, as sociologist, would introduce the concept of "rationality" in order to define the particular form of capitalist economic activity, bourgeois private law, and bureaucratic authority. And I have already discussed Parsons' alternative-value orientations in the introduction.

Equally significant for identifying the Western paradigm is the economic and technological system. Economically, the West believes it owes its success to the efficient operation of the free market system, and therefore to private enterprise. And I have already noted that Western technology is seen as the most advanced, supplanting all other modes of production that have preceded it, particularly since the Western invention of the art of invention itself: scientific technology.

Western medicine is identifiable with the unifying principle of allopathic medicine, whose philosophy proposes the treatment of illness by counteracting the symptoms of illness.[20] Other inventions that characterize the particular nature of Western societies include the educational system in which children, relieved of earlier functions, must spend long years of their life preparing to take their role in their society's life. The political system of the West favours the democratic consensus; the philosophical foundations of the West require that the individual be free to undertake his own purposes; psychologically, this involves allowing or encouraging the individual to develop his own personality to its full potential. This, in turn, is based on a deeper essential base in philosophical anthropology, where the individual is nurtured to claim for himself a historical identity, apart from those about him.[21]

The Marxist view can be closely identified with this framework. If the aim of science is to discover laws independent of human will and determinative of it, it follows that ascertaining those laws (that presumably determine human will) will allow those having technical knowledge of these laws to apply their knowledge in a technological way and to formulate the problem of social change as a technical problem. Here both the Western and the Soviet paradigms converge.

Much of the messages concerning Western culture came through the educational system and preached the view that local cultures were incompatible with the productive system the West had evolved. In the words of Ivan Illich, though not with his intent, the educational system, the media, and a host of other influences "schooled" members of the new nations in the dominant ideologies of the West; the means the West used, to take just two examples, education and medicine, were sought to be widely disseminated in the non-Western cultures.

Thus, indigenous medical systems were quickly derided and dismissed as quackery. And the idea that literacy was some sort of pre-condition for the acquisition of newer skills ignored some not-so-well-known facts about the industrial revolution: that men like Richard Arkwright learned to read and write in their old age. And who would have held ever that the Chinese revolution of 1949 and after required total literacy?

In the process, older forms of transmitting knowledge and skills were disintegrated. Take the case of the African, who is a member of a rich oral culture in which a three-dimensional diagram is preferable to the conventional plan and cross-section technical drawings, yet he has been forced though culturally unsuited and unprepared, to use technical drawings. This is a direct result of the idea that if the African worker is unfamiliar with any technical tradition, (which he decidely is not), the best thing to do is to change *him* to meet the needs of newer technical apparatuses. That these apparatuses themselves might have been modified to suit his needs has been ignored.

The Western paradigm, left to itself, might hardly have created problems. However, the West found itself facing a host of nations that in their constitution hardly approximated to its own image of itself. There were, also, the Communists. And, so, if in 1500 the European nations had begun a century of wonder on meeting Asia face to face, in 1950 and thereabouts Asia had begun and was nudged into a similar quarter-century of wonder as it observed the wealth, power, and technology of the West. In relation to that picture, the Southern nations found themselves, and were taught to find themselves, backward, traditional, underdeveloped, undeveloped, or developing societies.

The primary proposal of the West to the Southern countries eventually came through Walt Rostow: I have already commented on the real intent of Rostow's book, *The Stages of Economic Growth,* its ideological package, evident later in such crude policy terms as Rostow's positive contribution to the Vietnam War. Here I am concerned with Rostow's thesis itself, that the experience of the

Western nations with industrialization was being repeated by the Southern nations, unless they disrupted it by moving into the Communist stream. Rostow, for example, put the Indian "take-off" in 1952.

The Stages had no doubt an intellectual appeal difficult to resist: it contained the hope of discovering the laws of development, if not the philosopher's stone, a set of conditions which, if properly reproduced in the Southern nations, would bring about those processes which once led to the growth of the industrial nations.

According to Rostow, societies pass through five distinctive stages in the process of economic development: the traditional society, the pre-conditions for take-off, the take-off, the drive to maturity, and the age of high mass consumption. He emphasized further that these stages of growth are not merely descriptive but "have an inner logic and continuity. They have an analytic bone structure, rooted in a dynamic theory of production."[22] The presentation of the stages and the theory behind them are both outlined in the first sixteen pages of the book.

For the reader looking for any substantiation of the theory, there is of course none. There are not even specific criteria for the definition and the dating of the stages of growth. For example, "maturity" is said to be attained when new, modern techniques spread throughout the economy. But if "modern" refers to what is most advanced at the time, then neolithic Europe was a mature economy with techniques in advance of those of the old Stone Age. If, on the other hand, it refers to the adoption of techniques familiar to us now, this criterion of maturity is a piece of historical parochialism on the part of the observer whose criteria are confined to his own age: a point I have tried to make consistently throughout this book.

The most serious criticism of Rostow's theory flows from the nature of the world economy and how it historically developed. Rostow treats the progress of different economies mostly quite separately. Both the interdependence and the continuity of international economic development, detailed excellently by Woodruff in *Technology and Culture*, are obscured at important stages in the argument, which treats the development of the separate national economies largely in isolation.[23] As a result, the progress of industrialization is represented as the development of many disparate units instead of the interrelated process it was.

In fact, it is possible to show that the technological and economic development of the Western nations was intimately connected with special and non-recurring circumstances. If this is true, if the rapid increase in the material well-being of the West depended on novel elements (rather than upon ordinary and continuous and universally applicable elements), then a good deal of Rostow's ideas on the growth of nations is based on false assumptions and, therefore, false analogy. If history cannot repeat itself, future developments will arise out of new and ultimately unique circumstances. The Western past is unrepeatable and, therefore, Southern countries should no longer believe that their future can be determined according to the Western past.

What were indeed these special and non-recurring circumstances that facilitated the industrialization of the Western nations? Most of them have to do with the *extensive* phase of the Western nations, as they colonized old and new worlds. The industrialization of England, for example, cannot be understood without this experience that extended and facilitated its grasp of wealth and resources outside its boundaries.

I have tried to show how the mechanization of English industry was aided, against the power of India's handicraft industry, by political power. The exploitation of the colonies, on the other hand, was a convenience for the great European power, and I think critics, including Marxists, have had a hard time demonstrating that it was a necessity. The point, however, is that this convenience is no longer available to the Southern nations.

Another crucial element concerns the populations of the West, which after explosions found a safety outlet in the colonizing of newly settled areas: another facility not available to the Southern nations and their rising populations today. Population pressure on English land not merely induced a migration to the urban areas, but made migration from the country itself an economic necessity. International migration on a scale which is no longer possible contributed to raising labour's average productivity in the agriculture of the home countries.

Further, the colonization of underpopulated regions, primarily North America, but also South Africa, Australia and elsewhere by European settlers made possible the supply bases of European industrialization. And while capital obtained from the exploitation of the colonized (e.g., India) trading regions was transferred to Europe, capital was exported from Europe to the overseas regions settled by European migrants. The result was a sustained net transfer of resources during much of the nineteenth century from non-white colonies to white-settled underpopulated regions.

Thus, as the Indian economist, A.K. Bagchi, has concluded, the very processes that led to rápid industrialization in the capitalist countries and their overseas offshoots led to the stagnation or worse of the Southern countries. "This parasitic mode of growth in the nineteenth century makes capitalist growth non-replicable in underdeveloped countries of today."[24]

I have already rejected the view that the industrial revolution in England was a matter of conscious design, on the ground that such a view does not indeed explain why the majority of the English population had to go through a period of intense suffering and hard labour for at least a hundred years after things began to move: further, it should not be forgotten that this majority was able to attain a decent standard of living only after labour and trade union struggles. The myth that the machine brought about an automatic distribution of general prosperity is to be found in self-congratulatory textbooks.

I have also argued that England was in a period of ecological imbalance which forced its population to bring about a change in the existing resource base. The late eighteenth and early nineteenth centuries provided it comparatively easy outlets through which it could soften the transition. First, population migration eased the pressure on land: the migrants soon set about supplying England with raw materials (a possibility even the Chinese were excluded from). On the other hand, real colonial territories were opened up to produce raw materials like food: thus the transformation of the industrial nations from a state of rural, agricultural underdevelopment to one of the dominance of the industrial, urban sectors was only a seemingly real transformation. For, entire territories were opened up to produce coffee, bananas and a host of other commodities that the industrial nations needed no longer to grow on their own lands.[25]

When D. Ribero talked therefore of "traumatized" civilizations, and meant those whose lands had been taken up to feed a no-longer self-sufficient West, and those cultures that had been disrupted, their technics manhandled, their peoples maligned, I understood another kind of traumatization: that thousands of men and women from the urban centres of the Southern nations believed that in one way or another they could put a Rostowian programme of nation-building into action. If they continue to do so now, this is probably because they continue to go to Western school: it explains their dullness, and all their incapacity to think divergently. And school is not dead.

It is time to re-introduce what I tried to demonstrate in the first chapter of this book. I proposed there a new anthropological *model* to provide a new framework within which we found it easier to clarify the nature of man; I emphasized in that model, man's *capacity,* even *necessity* for dealing with his life-world through his two principal creations: technology and culture. Even in the chapter where we discussed "de-industrialization" I tried to stress the tremendous amount of adaptation that man proved capable of, an adaptation as revealing as the response man gave through the industrial revolution to problems that he was facing for the very first time in history, on a scale that required answers never before invented.

My model was an abstract model, so I "tested" it against the facts of history. In Latin America, Africa, Korea, the Islamic world, India, and China, thus, not only the Western world, we discovered various kinds of science (understood as systematic thinking about nature) and technologies created in the service of the members of these cultures. Different cultures, different goals, or different webs of meaning and, we saw this difference, even uniqueness, reflected and amplified in the nature of techniques.

The free ability to determine their own cultural programme was reinforced by their inborn technical capacity, which led in turn to a certain degree of independence concerning the relations of these societies among themselves, including the West. After the "traumatization" of the older civilizations, however, by the new found technological power of the West, (appearing, this latter, in response to ecological pressures) the non-Western capacity for technology and independent culture seemed to be devalued and underestimated by both sides, dominants and dependent.

By the 1950s, these civilizations and cultures consisted of leaders who had been taught or had taught themselves to accept the idea that they had nothing to offer of their own, that they must look to the outsider for ideas, for expertise, for capital, and *even for what they ought to think.*

The process of technological dominance was paralleled by feelings of cultural superiority. The anthropological *model* (to use the terms of this book) was de-universalized and assumed the flesh of the Western experience of human development: thus did the Western paradigm appear as *model,* even parade as one, and Western technology and culture as criteria for the understanding of other societies. All other paradigms were relegated to the heap of antiquated human experiences. It also seemed reasonable to argue that the goal towards which the world should be, or was moving to, was its "wholesale Westernization", Toynbee's image of the East succumbing to the West. The foundations of such a new world culture would, of course, indisputably be laid in the image of Western man.

What I have tried to do since that first chapter is to separate the model from the parochial paradigm to which it had become attached or with which it had turned out to be identified through the process of recent history. The divorce was necessary as a prelude to an argument to re-establish a plurality of paradigms, a result that historically would approximate, in terms of power, to a situation that existed before the disruptive impact of the Vasco da Gama epoch. Today, it might conceivably be possible to identify three competing paradigms, three independent paradigms: the Western, the Soviet, and the Chinese. No more. Such indeed is the poverty to which the mind of man has been reduced. As for the rest, we can only talk in terms of pale imitations.

Is the relativization of the Western paradigm inevitable or not? The future seemingly holds an ambiguous answer. On the one hand, as Greece and Chile show, the attempt to get out of the Western system or try out an alternative may be dis-established by the American or Soviet governments. And yet, it seems plausible to argue that the relativization is inevitable if we consider not merely one or two nations, but all the Southern nations moving out inexorably from total dependence to the industrial world.

The displacement of the West in its monopoly over the productive process will be accompanied by the displacement of its monopoly position as the arbiter of what is proper for the Southern nations in the realm of culture, ideas, and ideals. The wider dispersal of the ability to produce goods will be accompanied by the wider dispersal of the ability to produce ideas.

Let me argue the case for the distribution to a wider geographical area, particularly among the Southern nations, of the ability to produce goods.

In recent years, a new concept has suddenly entered into circulation among those who normally talk and write about the Great Divide between the North and the South: interdependence. It is a Western proposal, which promises "gains to all parties within a benevolent framework". There is talk of partnerships in "development", when anybody knows that "interdependence" among unequal partners can only result in the exploitation of the weaker partners. But let me examine the concept a bit more.[26]

At first glance there does seem to be a great deal of interdependence between the industrial and the Southern nations today. It is not disputable that exports of primary products – raw materials and other commodities – are crucial to the incomes of the new nations. At the present time, between 80 and 90 per cent of their export earnings are derived from primary products and more than 80 per cent of their foreign exchange accrues from exports to the industrial nations. There are some who feel that the international division of labour, which this implies, reflects an immutable law of economic life.

Thus, Daniel Patrick Moynihan could tell an absolutely incredulous Assembly that "if global progress in economic development falters", it would be the Southern countries that would submerge first. The economic health of the industrial countries, pontificated Kissinger, "is central to the health of the global economy". And another official went to the extent of claiming that "the most important single thing the OECD countries can do for the Fourth World is to continue to prosper".

This strange idea that the improvement of the condition of the people in the Southern countries must "remain a mere footnote to the prosperity of the developed world" seems to be a peculiar Western delusion. The arrangement of the world in which the Southern nations serve as sources of raw materials for Western industry and as export markets for Western manufactures is not one pre-ordained by any divine purpose. "It is true," noted the distinguished West-Indian economist, Sir W.A. Lewis, "that the prosperity of underdeveloped countries has in the past depended on what they could sell to the industrial countries, but there is no reason why this should continue."[27]

Lewis went on to assert that there was no reason on earth why the countries of Asia, Africa, and Latin America should not continue to survive and thrive, "*even if all the rest of the world were to sink into the sea*".[28] The simple reason for this blunt claim was, according to Lewis, the fact that the value of the total trade between the industrial and the Southern nations did not average even *four per cent* of the GNP of the industrial nations. As Carlos Diaz-Alejandro put it:

> The purely economic arguments pro and con openness in trade and finance have a way of becoming less than compelling when quantification is applied to them. Free trade enthusiasts are embarrassed by calculations showing that the gains from trade for most countries (North or South) typically come out

to be fractions of one per cent of Gross National Product when neo-classical tools are used in such estimations. They are put in the awkward dilemma of admitting that all the shouting is about peanuts. . . .[29]

Lewis also based his arguments on the fact that the free flow of trade and investment – at least in recent times – had done more harm than good to the Southern countries. On every ground, he observed, it is now possible to state that some of the Southern countries might be better off closing themselves off from the West. That it made better sense, for example, for the Southern countries not to compete with OECD nations in OECD markets, but to build up trade among themselves, an idea already proposed, but cut down by the IMF.[30]

More important, the Southern countries have all the resources needed for their own development, something that cannot be said, for example, of Europe and Japan. Taken together, they (Southern) have a surplus of fuel, fibres, iron ore, copper, bauxite and practically every other raw material. In agriculture, they are perfectly capable of feeding themselves, through exchange with each other: it is then perfectly ridiculous to beg the United States to buy more coffee and tea so that they can pay for American grain, when they could produce more grain for themselves.[31]

It was better, continued Lewis, to expand tropical trade with tropical countries, instead of with the temperate lands. Further, there was little sense in the Southern nations using their best lands to grow coffee, tea, cocoa, sugar, cotton, and rubber, "all of which are a drag on the market", when there was a booming trade or market for cereals, livestock, and other feeding stuffs. As for industrialization, it made again little sense to develop light manufactures – textiles, footwear, electronics, and the like – where competition is cut-throat and the market already rapidly saturated. There was no law of nature that laid down that the Southern nations must concentrate on light rather than heavy manufacture and import their machinery from the industrialized world instead of establishing their own metal-using industries "which is where employment is really to be found".

In other words, the real issues are not more votes at the World Bank, new rules for multinational corporations, reduced tariffs and better terms of trade, not even the stabilization of raw material prices in a constantly fluctuating market – though all of these might be seen as short-term objectives. On the contrary, the real issues concern the eventual industrialization of the Southern nations themselves, and thus a gradual turnabout from their present role as export economies.

The nature of these export economies is well known, has been long studied, and needs no fresh description.[31] But what would their industrialization, using their own resources, involve? What would be the consequences, for example, if together they succeeded in raising their share of world industrial oútput from its present level of 7 per cent to 25 per cent, as seems to be the determination all around? Such a programme, if put into effect by the year 2000, would actually wreck the absolute position of the West in the world's productive systems and relegate it to a relative place. How?

Conceivably, the increase from 7 to 25 per cent (which has been termed a "modest objective", it could be more) could be absorbed without dislocation, if we assumed a continuing high level – 8 per cent per annum or more – of world industrial growth. On current projections, the average annual growth rate may not rise more than 5 per cent. According to one calculation, this is barely more than is necessary just to keep unemployment in the United States down to its present inflated level.

Add to this, the increasing share of the Communist economies in the world market. The new programme, if implemented, would weaken the West's traditional hold over the world economy and drastically limit its ability to manipulate it in its own interest. In other words, the wheel which carried the West to pre-eminence in 1900 will have turned full circle by 2000.

The chances of this happening are all the more obvious when one sees that the Western nations are getting increasingly dependent on the Southern nations – not the other way round. C.J.L. Bertholet notes down the following two points (among others) that indicate a changing situation:

> In many developing countries, the industrial sector is growing fast, and in a number of cases, the percentage of the GNP accounted for by industry approaches that of the rich countries (the percentage of the population which works in industry is however generally lower). This growth is in any case much faster than in the rich capitalist world, and in a number of cases more than enough to compensate for the growth in population. A number of developing countries are getting much more control over their own resources of raw materials and fuel – to such an extent that the gradual shift of foreign intervention from the primary to the secondary production sector is certainly no longer due merely to the "pull factor" of higher profits but also to the "push factor" of growing power and self-confidence of the periphery. In fact, some authors go so far as to claim that the combined effect of these two factors in the future will probably lead to a gradual decay of neo-imperialist intervention.[33]

There is more evidence of this, and particularly of the multinational role in the entire situation, in the chapter significantly titled, *The Power of the Poor*, that magnificent volume on the power of the multinational corporations, *Global Reach*.[34] To be sure, as Barnet and Müller, the authors, point out the whole movement in that direction may be thwarted by new elements entering the scenario. And certainly, the West will fight back to keep the existing order intact.

This is the reason why the Southern nations have demanded in the Charter of Economic Rights and Duties, the "full permanent sovereignty" of every state over its "natural resources and economic activities", including the right to nationalize, expropriate, and transfer ownership whenever necessary to ensure effective control. It is in this light that the Indian nuclear explosion makes flawless political sense. Flawless if we remember Kissinger's surmise that "we are headed for an era in which economic problems and political challenges are solved by tests of strength".

The United States, Germany, and Japan have proclaimed it their task to continue the use of the existing international economic order, to improve and strengthen it in fact. The other European countries, on the other hand, seem to have come around to accepting the obvious. The whole problem, the French foreign Minister, M. Sauvagnargues, once declared, was whether the industrial nations were ready to accept "the principle and the consequences of accelerating industrialization" in the Southern nations. The Norwegian representative at the Assembly of Nations termed it the "challenge" all industrialized countries must face.

THE IMPORTANCE OF THE GENERATION OF SKILLS

The Chinese seem to have already provided an excellent example the direction in which the Southern countries may move: I am talking here of the pattern of industrialization adopted. The Chinese have been aided in this by the influence of their own historical past.[35] The isolationism of Chinese thinking is a quality of Imperial China, which was scarcely confronted by the problem of "things foreign" as a result of its own assumed cultural superiority in most fields, which led in its turn to presumptions of self-sufficiency.

Mao's commitment to a mode to technological development that did not attempt to imitate industrialization as it exists in its present form, but to form a base of industrial activity in spheres where it became the leading edge of agriculture, was not, of course, the result of a policy set forth from the Chinese Revolution of 1949; at most, it was after a great deal of trial and error and the Soviet experience of 1960 that the present strategy was concretized. The Southern nations are probably fifteen years late, but their inclination for a more autonomous technological development seems to stem from a similar disenchantment with foreigners, this time the Western nations. But the Soviets at least were committed to the industrialization of China, something that cannot be said of the relationship between the Western nations and those of the South. So it should not be long before the Southern nations adopt the policy that foreign trade should only be a supplement to basic internal development of their resources.[36]

The vast size of the Chinese domestic market and the availability of most of the important raw materials within China means that Chinese economists do not have to seek foreign contacts, a conclusion that should be equally obvious and applicable to the Southern nations should they act in concert. One cannot be realistic if one expects the Southern nations to learn their valuable lessons and change overnight: the important thing is that Southern nation economists and planners do not think as they did twenty-five years ago, which signals a gradual, but thoroughgoing revolution in consciousness about strategies, most of the newer ones proposed being in fact increasingly divorced from the Western paradigm.

Industrialization and its acceptance as a leading factor in the economies of the Southern nations is no longer a matter of dispute: most of these nations at least seem to think it necessary, a minimum amount of it, and if this decision has been made with a view to the circumstances of their cases, and not to continue on their race with the West, and if we grant the relevant leaders concerned, the freedom to make their decision about it, there is very little to say, except perhaps indicate how the Southern economies might just about go into it the surest possible way: here the experience of China is invaluable.

An engineer who is also a craftsman will easily understand what follows. The transition from "manufactures" to "modern industry", as Arnold Pacey has so well brought out, may have meant abandoning traditional skills, yet the relatively gradual and continuous transition which was effected in Europe meant that some knowledge, skill, and experience was usefully transferred from the older form of industry to the new. In India, on the other hand, or even Tanzania, some of the older forms have been destroyed either in colonial times or *even under present governments*, or they have been left unattended, while new ones have then been tried out unnecessarily on barren soil.[37] Michael Polanyi has beautifully analyzed the meaning of craftsmanship elsewhere, and so has Ortega y Gasset in the essay I have discussed earlier, but here the Chinese approach to the problem might just as well suffice. Perhaps "approach" is the wrong term: what "happened" in the Chinese capital goods industry seems more to the point.[38]

The Chinese capital goods industry is presently dominated by two distinct groups of firms: one, the recipients of major investment inputs and Soviet equipment imports during the 1950s, including a few with pre-1949 histories of Japanese ownership; the other group includes smaller enterprises, but these are much older and constitute the legacy of pre-war industrial activity in the private sector. This latter group received relatively minor infusions of investment or imported equipment after 1949.

In the 1960s, however, a decade in which economic difficulties and international isolation forced domestic producers into larger attempts to improve quality, variety, and manufacturing techniques, it was the second "traditional" group that came to the rescue and established a clear position of leadership. Why? The simple answer might be that this group could substitute experience for investment and thus indicate and stress the importance of the accumulation of skills in enlarging China's production possibilities. There is nothing that can substitute for skill development, however, as the Russian-aided plants showed. The second group's accumulation of their technical skills was due to a remarkable period of continuity, evident in this history, for example, of a firm like the Talung Machinery Works of Shanghai.

These works were begun in 1902, and principally employed *repair* specialists, who later found it easy to move on successfully into textile machine repairs, parts manufacture, and only after all this, into the production of complete sets of cotton-spinning equipment. It was the skills acquired by the Talung veterans and other private firms, and by those who had studied abroad or worked in railway shops or foreign-owned factories that enabled a rapid growth and diversification of the entire machine industry during the final pre-war decade (1927-1937). It was the presence of these skills that made the transition from repair work to manufacture of complete equipment less difficult that it would have been otherwise.

The new government of 1949, however, ignored these older and smaller units during the first decade, and instead concentrated nearly half of overall industrial investment in 145 Soviet-aided projects. With few exceptions, these were either created anew, such as the Loyang complex, or expanded Japanese-built plants, such as the Anshan Steel Works, in which earlier Chinese participation had been limited to supplying semi-skilled labour. Shanghai, the centre of China's older, inherited industry, which turned out 22 per cent of machinery output and 19 per cent of national industrial output in 1957, received only three of the Soviet-aided projects, and a mere 2.5 per cent of overall investment during 1953-57.

China's second Five Year Plan (1958-62) continued to be drafted with the expectation of continued large-scale acquisition of Soviet blueprints, equipment, and technical expertise, but was disrupted by the sudden withdrawal of Soviet technicians in 1960, of Soviet supplies of petroleum, military goods and equipment needed to complete unfinished construction projects.

It was precisely the enterprises and regions that had received little investment during the fifties that could now rise to fill the gap. The petroleum industry, which has achieved a tenfold output increase since 1960 is equipped by primarily converted engineering plants including former textile machine repair works like the Talung Machinery Works mentioned already, and the old Lanchou General Machine Works, a provincially managed descendant of a nineteenth-century *arsenal.* The Talung Works, which have now pioneered in manufactured petroleum equipment, compressors, and machinery for producing artificial diamonds, illustrates the continuing innovative success of the older engineering firms.

Capital investment cannot be the key to the explanation why these older and smaller firms have proved readily capable of developing new products and techniques: we have observed that investment bypassed them in the fifties. Their strength is a matter of technical experience, which is essentially a long-term affair. In other words, China seems to have gone through the first industrial revolution in its entirety.

There is no way in which the Southern economies can bypass this crucial stage. In fact, unless this first stage is gone through, the second and third industrial revolutions will never see an integrated and dispersive development: there is no way to short-cut the process by going directly to the higher level stages. In spite of all that the industrial nations may advise, the most profitable kind of technology they have to sell belongs to those higher stages. This is why a lot of what goes under the label of "transfers of technology" is maladaptive.

The import of foreign capital goods and equipment may be the quickest and cheapest method of injecting new activities into a pre-industrial economy, but its long-term effects are negative, *as it curtails the skill formation process*. It also prevents domestic firms from learning to implement new techniques without purchasing equipment. Should exchange problems or political trouble disrupt foreign equipment supplies, domestic capacity to absorb new methods suddenly becomes a constraint on the productive system as a whole.

In China, the sequence of repair, parts manufacture, and full manufacture is encouraged in both urban and rural areas. Former *repair* shops have begun to manufacture trucks, locomotives, and agricultural equipment. Small rural producers of fertilizer, cement, iron, power, coal, and machinery are expected to repair and often even to manufacture their own equipment.

Thus, the accumulation of skills has also direct relevance in the industrialization of the rural areas, which presupposes the diffusion of industrial skills to large masses of the Chinese population. A clear example here is the Chinese adoption of the use of vertical shaft kiln technology in most small-scale cement plants. The number of these plants increased from about 200 in 1965 to 2,800 in 1973, and total output increased from roughly 5 million to an estimated 20 million tons.

Jon Sigurdson thinks one of the reasons for the speedy adoption of the small-scale plants is probably the high transportation costs involved in transporting cement from large-scale plants using rotating kilns. But he is also keen to note the impact of these plants on the diffusion of skills:

> A large number of people are being trained in industrial process technology. A sizeable number of people have, inside production units, received training in organizational skills. A smaller number, but still sizeable, have been trained in administrative skills related to the procurement of machinery and raw materials, distribution of products and coordination with other industrial units.[39]

India seems to have moved in the opposite direction. Indian experience with vertical shaft kiln technology thus far has concerned four plants: two have failed. The development of these plants on a larger scale was stalled when larger manufacturers intervened: it is well known that at least in the sixties as the movement might have been taken seriously, that the cement adviser to the Government of

India also happened to represent a large Danish cement machinery manufacturing firm in India.[40] Further, it should not be overlooked that many of the proven existing mineral resources for the manufacture of cement have been leased out to established large-scale cement manufacturers, who have obtained licences for their activities. And it is in the cement small-scale industry that the ITDG (London) begins its counter-local expertise movement.

Perhaps, it is possible to argue that the transfer of technology through foreign firms might at least lead to a diffusion of skills concerning modern large-scale processes. The evidence, say Michael Kidron, Barnet and Müller, points the other way: Kidron writes:

> Research and development are invariably conducted abroad; the fruits of development are invariably imparted, if at all, at very high cost in royalties, fees and other payments, and not always in their entirety; through their production and staffing policies the major investing firms attempt to systematize a continuing control of know-how; and much else in the same vein. Since the Indian partner is normally assigned – and readily accepts – a narrowly specialized range of functions, the diffusion of skills that does take place is largely fortuitous. Indeed, since the typical modern investing firm owes its dominance and income largely to its technological monopoly, a different outcome would be surprising.[41]

Sales of techniques or know-how come through collaboration agreements, signed between Indian industrialists and foreign companies and the majority of them are dictated more by the Indian industrialists' eagerness to exploit Western technology for quick profit than for any other more meaningful purpose, and more often than not they are meant for the production of non-essential goods: vacuum flasks, lip-sticks, toothpaste, cosmetics, brassieres, ice-cream, gin, chocolates, beer, biscuits, dry batteries, ready-made garments.[42] All these and more are high-income wants in India.

The situation within the public sector managed firms is hardly different. Who is in charge of the Indian Planning Commission or Indian planning? Evidently Indians. In reality, foreign experts supplied by international corporations through private or governmental channels. Multi-purpose river valley projects, atomic power plants, flood control programmes, agricultural research, geological surveys of mineral and oil resources, town planning, railway expansion, road development, bridge construction, the manufacture of indigenous tractors – in each of these areas foreign technical experts have been *preferred* to Indian engineers.[43] The latest instance of this unfortunate xenophilia is the construction of the second Hooghly Bridge near Calcutta.

The project report was initially broached by an organization created by the Ford Foundation, and approved by the Foundation's traffic experts. The World Bank prepared to finance half the construction expenditure, evidently since the bridge would enable easier handling of raw materials, especially minerals from the north-eastern belt to the Calcutta port for eventual processing by multinationals. The techno-economic survey was done by Rendel, Palmer, and Triton of London. Finally Freeman, Fox, and Patterson of London were asked to design the bridge. Germany's LUNA concern was also invited to join in. The fees payable to all these experts: 10 million rupees. The Indian engineer was ignored, not because he has no experience, but because foreign aid for the bridge came tied with the stipulation that foreign expertise be used.

If India is in some sense representative of Asia generally, Nigeria may be regarded as representative for Africa in a similar context. Nigeria is one of the more active of the African economies in terms of the investment in technical training and the quantity of trained manpower available. Here, too, however, most of the training seems to be not properly utilized. The Nigerian Society of Engineers placed the following advertisement in local papers on September 11, 1970:

> The Society is amazed that in a country with two or three Universities producing graduate engineers, most of whom have not been able to secure employment in the engineering field, government officials should continually feel that our development projects can only be conceived and executed by foreign engineers.... For instance, the Kainji Dam, the largest engineering project ever undertaken in Nigeria, produced no benefit to the country in engineering manpower development.... If Nigerian graduate engineers cannot get employment on engineering projects in Nigeria, how are they to obtain their practical training and experience? We note that the various governments [that is, of Nigeria] are obsessed with the recruitment of foreign engineers.... The Society believes that the whole purpose of University education and manpower development will be defeated if concerted plans and legislations are not made to enable young engineering graduates to be assimilated into the economy of the nation by way of "on the job training". The Nigerian Society of Engineers has been endeavouring to get people in government circles to make it obligatory for foreign engineering firms who control the bulk of the consulting, construction and manufacturing industries in this country to employ Nigerian engineering graduates....[44]

Contrast the Indian and the Nigerian cases with the one of the former secessionist enclave of Biafra: what happened with regard to this area's technological competence can be shown to have occurred in similar fashion in the technological histories of Europe (as distinct from Britain), Japan, and America. The Chinese, after the Soviet withdrawal, provide one more example.

I am thinking here of one of the possible consequences of delinking one's economy from the rest of the outside world.[45] As Carlos Diaz-Alejandro observes, in the first place, delinking will not automatically lead to better conditions for the de-linker, as the examples of Burma and Tibet (before Chinese incorporation) show. But there is a great deal of significance, as he shows, in making the world safe for selectivity: that is, enabling countries, particularly Southern ones, to establish selective links with the North. Biafra is an extreme case (not even China comes close) of what can happen when an economy is de-linked totally from the outside world.

During the Nigerian civil war, Biafra found itself blockaded from the outside world, thus forcing its engineers and technologists to go to work on their own. The result needs mere description and no comment. A broad range of consumer and non-consumer goods were soon indigenously produced, all in small-scale units, and at a fraction of the capital/output of equivalent installations in Europe or in the rest of Nigeria at the time. Goods produced included not only petrol and diesel fuels, but soaps, engine oil, protein extracts and salt, besides vaselines, chalks, and biscuits.

Pure salt was produced with small locally-fashioned units at a cost no more than £3,000 and capable of a little over ten tons a month. A refinery, the result of a wide range of adaptations and innovations, and capable of producing petrol, diesel, and kerosene quantities of 15,000 tons a year was set up at a cost of £50,000. The mechanical parts were fabricated and welded on site. The high capacity furnace of the normal high-scale, sophisticated refineries was replaced by very simple but effective home-made burners. Compared with a refinery built in Nigeria by Shell-BP, this unit gained as much as 400-500 per cent in capital/output. A large number of home-made "cooking-pot" refineries were begun to produce crude petrol at costs as low as £300.

A soap factory at a cost of £25,000 was producing as good a product as a £1 million, foreign-owned soap unit in Algeria. The plan to produce cement in small-scale units was ultimately disrupted by the advance of the Federal forces. This inverse correlation between the absence of foreign technical aid and the spurt in indigenous productive capacities is available even in the case of some Indian industries.

Early in 1924, the colonial government of India imposed a small tariff on the foreign match-making industry in the country. The tariff was introduced to protect the indigenous match-making industry. The Swedish Match Company, against whom the tariff had been directed, changed tactics, and began to manufacture matches through two new subsidiaries, with Indian names: the Western India Match Company and the Assam Match Company. By 1945, its eleven factories in both subsidiaries were supplying 80-85 per cent of demand. Further, it even continued to provide the necessary raw materials, including wood, and chemicals, to the indigenous industry itself. The small-scale units, mostly cottage-based, continued to disintegrate, until the government, after independence, intervened to restrict the output of the Swedish units.

The consequences were soon obvious: cottage industry expanded its production *three and a half times* between 1949 and 1961. (The issue still remains, of course, basically unsolved, for Swedish match continues to produce three-fifths of total output, and still controls nine-tenths of the production of potassium chlorate, one of the principal ingredients in the manufacture of matches.)

The second example concerns Unilever, with an Indian subsidiary operating under the name of Hindustan Lever. This particular firm today holds the commanding position in the soap and detergents industry. Immediately preceding the Second World War, it was producing nearly one-third of total factory output. It cornered 70 per cent of the market after soap rationing was ended in September 1950, with the result that the largest indigenous producers were operating at one-half to one-third of installed capacity compared to 94 per cent at Lever. In 1960, Lever was still selling 83,000 of the 152,000 tons produced, the decrease having come about when Lever realized that it would be better to cut down production than to further antagonize indigenous producers. The cut-down resulted in the rise of the indigenous sector itself.[46]

To most development experts or elite planners in the Southern nations, any process that does not use "advanced technology" techniques is at least theoretically out of date and will soon be supplanted. The Biafran experience was belittled, for example, as the octane value of the petrol produced in the small units was 80-85, thus not as efficient as the "advanced units". If we consider the number of years, however, it took European and American refineries to achieve their present octane values, it then becomes easily apparent that the Biafrans, in obtaining such high octane values in less than twenty-four *months,* were simply very good engineers indeed. E.F. Schumacher would have called it all "intermediate technology".

The real question, however, is whether any technological system is more advanced than another. If the term "advanced" is used merely to refer to a technological system suited for areas of high level economic status, there is then no basis for any attitude which seems to suggest that advanced technology is any more dignified or complicated than technology suited for other areas. The skills and ingenuity of engineers and technologists are needed in each economic area and the task is no easier whether the area is economically advanced or not. Every technological system requires the same level and amount of resourcefulness and originality.

And it is also obvious that in the measure in which the Southern nations increase their dependence on foreign technological systems, they will also proportionately diminish the technical abilities of their own peoples.

THE RELATIVIZATION OF CULTURE

This brings us to the other principal element of our *homo faber* model: culture. For, once the monopoly of the West over the production of goods is restricted, the event will create new and legitimate problems at the level of the production of ideas. The world of learning will never be the same again, having entered a new phase, where the question of what constitutes knowledge and history will find as many different answers as there are cultures.

In their great eagerness once to "catch up" with the Western nations, the Southern countries I observed tried to take over the *means* used originally by the industrial nations. What is not so easily realized is they also took over the goals of the societies or models they set out to imitate, which resulted in a devaluation of faith in their own values, culture, and civilization.

To put it differently, a new normative structure was sought to be inculcated, and more often than not, this structure of norms went directly against the grain of what the Southern nations had inherited from their past. This resulted in disrupting their possibility of preserving their older cultures *and* their ability to create newer adaptations in continuity with the old. Once the normative ideals of the West were adopted, or seen as superior, it became necessary to force people to adapt to the allegedly superior culture.

Wrote the Turk, Yasar Kemal, once:

> I do not believe that most Turkish intellectuals with all their good will, will be helpful to socialism. For 200 years the Turkish intellectual has aped the West; imitated the West. An ape is not creative. It may look human but it is not creative. Since Turkish bourgeois intellectuals have aped the West for 200 years, they have not made any contribution to Humanity for 200 years. . . . Now after 200 years, when we say let us turn to our Identity, let us set up our own socialism, real socialism, then they turn against us, and start looking for models. . . . They look for the model of the Soviet Union or the model of China. The entire struggle of Lenin has been to create his own model.

And that other Islamicist, Mehmet Akif, rounded up the final consequences of all this imitation fever with the following choice words:

> People of a nation, whose religion is imitation, whose world is imitation, whose customs are imitation, whose dress is imitation, whose greetings and language is imitation, in short, whose everything is imitation are clearly themselves mere imitation human beings, and can on no account make up a social group and hence cannot survive.

Gladly, we are entering now a new period, characterized principally by the realization that not only was our imitation of the West and its culture our psychological problem, but that there is little any more in Western culture or even in the intellectual presentation of the Western tradition that is any more touched by "absolute" values. The idea for example that the West is out to preserve the world and keep it *safe* for democracy has been exploded to smithereens. The irony surrounding the sixty-strong CIA team in Athens that considered it more important that Greece could be a bastion against communism than that its democratic government should be preserved (in the very birthplace of democracy) is further sharpened with the following quote from a letter of the late President L. Johnson to the Greek ambassador, who protested the American plan:

> Listen to me, Mr. Ambassador. Fuck your Parliament and your Constitution. America is an elephant, Cyprus is a flea. Greece is a flea. If those two

fleas continue itching the elephant they may just get whacked by the elephant's trunk, whacked good.[47]

Recent decades have produced research that question the "democratic" functioning of most of the Western industrial nations.[48] Every aspect of their societies, including the medical, have in recent years come under heavy criticism and disillusionment, principally from Westerners themselves.[49] It remains to say something finally about the Western educational system, and the Western academic world.

It should be clear by now that the Western academic presentation of the world has run into serious difficulties, most of them a result of ethnocentric bias. There has, indeed, been very little of objective scholarship in citadels of knowledge that once claimed to be the arbiters of objectivity. I have tried to indicate these grave inadequacies in numerous sections of this book. I do not know about the education of Western children, but certainly it would be a great error to permit the education of non-Western children with these ethnocentric works.

In the history of technology, at least most *written* histories of technology, I have shown for example that a great deal of evidence has been illegitimately ignored, twisted, and wrongly interpreted, giving the final impression that there has been but one essentially worthy development of the mode of production, the current Western one. If others have existed, they have been treated as footnotes.

A theory that does not take into consideration as many facts as are available is generally unacceptable. It is time to propose that a work that deals with the history of technology, but which is actually a history of the technology of a part of the world, is certainly no real history of technology, and should not parade under that name. This proposal should be seen to extend to every other area of academic life, whether we are discussing art, ethics, or childhood.[50]

No longer is it possible to simply assume Western objectivity in scholarly matters. Worse, there is a growing body of evidence that indicates that Western scholars have cooperated sometimes with their rulers in the latter's bid to further and continue their exploitation of the people of the Southern nations.[51] There is therefore need for a list of all such scholars that have cooperated in such tasks. Ultimately, there will come a time when every Western work betraying an ethnocentric bias will have to be removed from the libraries and committed to the flames, in the words of David Hume, "for it can contain nothing but sophistry and illusion".[52] New books will have to be written for a more universal age no longer carved in the image of Western or European man.

THE THIRD LEVEL: THE OTHER HALF THAT DIED

I have repeatedly turned to China in this chapter, not because I think that the Chinese have answers to all the problems the contemporary world faces; nowhere, indeed, would I wish to propose the replication of the Chinese paradigm *ad nauseam*. Neither the Chinese nor the West has any monopoly about the way the world should be; or, in the long run, about how life may be lived.

What I do appreciate about the Chinese paradigm is its revolutionary perspective about the role of people in the evolution of a society and its goals. There have been in history two major approaches to this issue: one has believed in direction from the top, the other, participation from below. In the latter case, participation from below, we seem to come quite near to the horizontalization of the members of a society. China *tends* to the second alternative, and I tend to it too.[52] Roland Berger once emphasized the very early nature of the Chinese tendency:

> In his August 1955 speech Mao Tse-tung used the phrase *tzu li kong sheng* which literally translated is "regeneration through our own efforts". This more accurately conveys the true meaning of the policy than the term "self-reliance". "Regeneration through our own efforts" also makes it clear that this is a policy radically different from "self-sufficiency" or "autarchy". It is in fact the mass line applied on the economic front and stems directly from Mao Tse-tung's consistent emphasis that "the people, and the people alone, are the motive force in the making of world history" and that "the masses have boundless creative power".

Except perhaps in a few other countries like Cuba and Tanzania, the Chinese elite's attitude to the rest of the Chinese population has no real counterpart in any other region of the world. I have here in mind not merely the industrial nations, but more especially, the Southern countries. China has, more than any other society, released the creative capacities of its people. Within the space of a mere twenty-five years, it has passed through its industrial revolution with little of the suffering that accompanied the revolutions of the Western nations.

Let me take merely the Southern countries: when men speak of the "poor" countries or the "Third World", they normally do not qualify their statements by observing that there are rich people and poor people in this region. It is, in fact, more than evident that these rich groups have appropriated most of the benefits of "development" programmes over the past twenty-five years, that they are the key channels for the entry of multinational corporations into local economies, and that more often than not, they use the power of Western arms to repress their own populations. Nobody in his right senses would be willing to deny that these privileged groups have grown richer in the recent past, and that their poorer compatriots have grown poorer.

I have therefore consciously evaded, throughout this book, the proposing of any facile solutions to the poverty of people in these countries. I do believe in fact that the most beautiful blueprints for the eradication of poverty have long since been proposed and that they cannot be expected to make any dent into the problem so long as the elites that control these countries do not want any solution to the problem.

So we have the Norwegian philosopher, Johan Galtung, producing scintillating pieces on "self-reliance", or a Myrdal with a world anti-poverty programme. Rarely do we get a Diaz-Alejandro, who after studying the merits and demerits of a proposal such as delinking, or selective linking, would note:

> So even if all rules allowing for selectivity in international markets were accepted, or indeed, even if all Southern demands now embodied in calls for NIEO were to be immediately accepted by the North, it is doubtful that the mass of the poorest citizens in the South would feel much improvement in their welfare at least in the immediate future. This is just another way of saying that the central development problems for most LDCs are internal.[54]

There are those who feel that better commodity agreements and stabilized prices for raw materials for coffee and tea will help the people of the Southern countries. They do not wish to realize that better prices will merely aid the minorities who own the best land in the Southern countries and who produce these cash crops to get richer still, and to thus aggravate the disparities in income levels in the region.[55] And oppression is a natural consequence of very wide disparities in income levels.

The greatest amount of oppression takes place within the Southern nations, where the disparities are largest. There are no poor people in these regions, only the oppressed. This oppression has a political basis, and no amount of intermediate technology or delinking moves will ever dissolve that base. Thirty years after independence, lower income groups in India like the Harijans still get their houses burnt, their wives raped, and themselves murdered if they attempt any moves at emancipation or organization for social justice. There are still people naive enough to think that a gas-plant or a new sickle will solve the problem. They won't.

There are those who also believe quite sincerely that extending the arm of the multinationals into the Southern countries will bring the latter more aid and technical expertise to help them solve their problems.[56] This is actually a view of multinational corporation spokesmen themselves. Barnet and Müller have studied the problem in minute detail and come up with a host of negative ideas. They note:

> The unfortunate role of the global corporation in maintaining and increasing poverty around the world is due primarily to the dismal reality that global corporations and poor countries have different, indeed conflicting, interests, priorities and needs. This is a reality that many officials of underdeveloped countries lacking alternative development strategies, prefer not to face.[57]

Barnet and Müller see a different scenario for the future, when intense competition among the industrial giants for scarce raw materials, "export platforms" (low-cost labour enclaves from which to export to the industrialized world) and new markets will give the Southern nations an importance in global industrial planning that Lenin prematurely proposed for it in 1913. This could only mean the further extension of these corporate giants on the soils of the Southern nations. In the past, every such extension has led to more oppression of lower-income groups.[58]

In *How the Other Half Dies* Susan George provides some frightening evidence of how this happens.[59] Most of it takes place in the context of the New Agricultural Strategy, or the Green Revolution popularly named, as land is appropriated from peasants to build agricultural universities, or these same peasants are driven into the cities as huge dams flood their holdings forever. Farm mechanization itself is reducing the need for farm labour just when we need to have more people employed.[60]

My final point is that most of all this is probably quite well known. The studies on the Green Revolution, on the pernicious effects of multinational corporations and the abuses of corporate power in the industrial nations themselves, much less the Southern nations, are legion. There are any number of studies on the poor, and this is my principal complaint. For, in any context in which the poor are oppressed by elites, any further information will always be used against them. It seems reasonable to propose a moratorium on all studies on the oppressed.

One of the more courageous statements in this regard is contained in the speech of Martin Nicolaus to the 1968 convention of the American Sociological Association, which bears repeating even at this late date: He said:

> The corporate rulers of this society would not be spending as much money as they do for knowledge, if knowledge did not confer power. So far, sociologists have been schlepping this knowledge that confers power along a one-way chain, taking knowledge from the people, giving knowledge to the rulers.
>
> What if that machinery were reversed? What if the habits, problems, secrets and unconscious motivations of the wealthy and powerful were daily scrutinized by a thousand systematic researchers, were hourly pried into, analyzed and cross-referenced, tabulated and published in a hundred inexpensive mass-circulation journals and written so even the fifteen-year old high school drop-out could understand it and predict the actions of this landlord, manipulate and control *him*?
>
> Would the war in Vietnam have been possible if the structure, function and motion of the U.S. imperial establishment had been a matter of detailed public knowledge ten years ago?

This is indeed a good programme for future research and study on a world-wide scale. For the past many centuries, we have had too many questions asked about *us,* about the oppressed. It is now time to reverse the tables, to enquire, for example, into why America wets its pants every time communism is mentioned; into why the insecurity complexes of the Pentagon have proved so long-lasting; to ask why nations that trace their ancestry to democratic Greece in text books have in recent years not merely toppled democratic regimes but continue to support totalitarian regimes elsewhere; how nations that claim to have produced the Renaissance can spend so much time, energy, and thought in the almost paranoid pursuit of profit; how nations that claim to be democracies continue to cloak their largest industry, the production of arms, from public scrutiny and democratic control. Finally, it is necessary to ask how nations with more than forty per cent of their best scientists and engineers engaged in the production of weapons to destroy human lives all-round the world can advise the Southern nations to use "appropriate technology".[58]

THE LOGIC OF APPROPRIATE TECHNOLOGY

It seems more than obvious that the further one moves along a continuum from poor to rich the more inappropriate the technological system tends to become. At the lowest levels, the technological system is already appropriate or intermediate as the case may be. The poor and oppressed have always had to make do with restricted means. They have also been more friendly ecologically. I spoke early of the survival engineer engaged in the preservation of his life in a hostile human environment.

At the other side of the continuum is the engineer engaged in a reverse ambition: the destruction of life: the two types of engineers seem to be meeting already as the Western world and that other megamachine, the Soviet Union, rush arms to Southern governments increasingly resorting to violence to quell the despair of their hungry peoples. It seems as though the engineer from the industrial nations and the survival technician, both human beings, can only meet as dogs of war.

The industrial engineers from the Western world work in a system, it is generally admitted that wastes a great deal of the resources of the globe. The Club of Rome report on the predicament of mankind is not meant for the consumption of the Southern nations for these do not waste resources to any extent as those of the North. And when *Small is Beautiful* appeared, few people realized that all the chapters except one dealt with the inappropriate technological system of the West.

It was the Dutch engineer, Ben van Bronckhorst, who observed that the changed circumstances involving raw materials from the Southern countries and the energy crisis made inevitable a number of changes in the technological system of the West itself. He wrote:

> The search for new ways is not restricted to the countries of the third world but certainly involves the rich countries themselves. It is essential to indicate case by case in what direction the solution must be looked for. . . . This has given the discussion of techniques a new shape: the issue is no longer the contrast between modern and alternative technology, but the necessary technological development needed in each country.[62]

The relevant question here, however, is who is going to help bring about this shift to the necessary and desired technological development for each country. What is involved here is the issue of the distribution of power, and the concentration of it in the hands of private interests and concerns. The politics of technology often turn out to reinforce the *status quo,* and the question remains: who is going to impose solutions and where are the politics to promote their imposition? How do we help separate the aims of a democratic society from those of private industry, for example and make the former control the latter?

There is a danger if issues are not confronted in these terms. Take the environmentalist movement in the United States: a massive bureaucracy has been created at the federal level for the purpose, so massive that there is a danger that only private companies with their immense resources could deal with such a bureaucracy, and eventually succeed in capturing it when it had been set up to supervise *them.* For, as it may turn out, only a large corporation would have the finances to manage the economists, lawyers, lobbyists, tax experts and others necessary to guide it through regulatory obstructions. This could conceivably lead to a further refinement of the system under which corporate power dominates the state.[63]

It is precisely such a situation that links the conditions of the oppressed in the Southern nations with that of the unemployed and other concerned groups in the Western nations, particularly those keen on confronting corporate and other vested interests in order to bring about a more human world and a gentler technology. The number of Westerners who have come to regret that their societies ever placed their economies higher than man himself is not to be underestimated.

The followers of Marcuse might claim that the proletariat of the industrial countries have had their fires defused and been accommodated into the attractive lure of a capitalist-induced, consumer economy, and they may in fact be right. But such an occurrence, hypothetical or real, has happily been followed by the rise of militant non-poverished groups. The Dutch contribution to the victory of Vietnam over the United States is well known. So, by now, is the famous Nestle case, where a militant group based in Berne, ultimately paid a fine of 300 Swiss francs to devastate the reputation and force changes in the advertising tactics of that multinational corporation. Here, it is easily seen how interests across nations and cultures coincide, and can form the basis for further cooperation in the coming struggle to tame the power of the multinationals.

Afterword (1991)

500 years ago a Spaniard named Christopher Columbus sailed out from Europe to set foot on a piece of earth his countrymen had never seen before. This book is being brought out by the Other India Press in a special edition to coincide with the fifth centennial of that epochal voyage. It also has a new title, the credit for which goes to Ward Morehouse. The original title (*Homo Faber*) was intelligible, it now appears, mainly to Roman Catholic priests and seminarians.

Though I wrote the book as a doctoral dissertation in 1976, how aptly it has come to suit the new occasion! *Decolonizing History* is concerned precisely with the impact of the five centuries of interaction and colonial dominance on the way we perceived nature, ourselves, others. What a splendid opportunity to regurgitate the book's themes! This edition is therefore being issued without any changes.

The Columbus voyage was not only a daring achievement, it set in motion a deadly chain of consequences that radically transformed the planet. No other journey undertaken either earlier or in any other part of the globe has quite compared with it to this day.

Within six years of it, the Portuguese navigator Vasco da Gama landed at Calicut on India's west coast, inaugurating what K.M. Panikkar would call "the Vasco-da-Gama epoch". A decade later, in 1510, Alphonso Albuquerque conquered Goa (where this writer lives). As a consequence of those fateful journeys the destinies of Europe, South America, Africa and Asia were soon intermeshed.

The invasion sowed the seeds of the idea of Europe, a collective notion identifying 'us' against all those 'non-Europeans'. The invention of Europe proceeded apace with the 'discovery' of the new world. Europe's contrasts with the latter were emphasized so that it could gain an identity it lacked, a security it sought, problems faced neither by the Indians or the Chinese securely rooted in civilizations with prodigious survival power.

The balance sheet of the 500-year encounter between the West and our part of the world is yet to be drawn up and there are strong opinions being expressed by cultural auditors on either side. For the Europeans, the 5th centennial is an occasion for pomp and celebration. For the people inhabiting the worlds allegedly "discovered" by Columbus, da Gama and Albuquerque, the memory of the past five centuries, being largely years of trauma and violence, is better forgotten. The Africans, in fact, are demanding reparations, while the South Americans are insisting that what occurred during 1492 and thereafter cannot be glossed over as an 'encounter' between two worlds. It was an invasion coupled with a violent conquest and takeover.

This is true. However, it still remains only a part of the story of the five centuries now past. For in addition to the invasion of South America and Asia, Europe also invaded itself and radically altered its own soul. Because of a series of well known technological and socio-political transformations, Europeans arrived at a belief that there were no limits to be encountered to the rational manipulation of either man or nature. Thus this grand urge to re-make the people of other worlds in the image of Europe was not an incidental quest. It was an intimate part of Europe's express desire to subordinate human societies from every part of the known world, including its own, to the new project of inventing a new kind of man.

Technological prowess led to the assumption of moral superiority as well. Only Europe mattered now: every other civilization was seen either as a crude prelude to Europe or of no further relevance. That other civilizations might have other programmes was deemed interesting but of little significance. Europe decided its values were the only authentic ones: its art, music, architecture, ideas, social organization, its ethnicity in short, would become the only universal for all men, and world history written to fit these assumptions. The socio-geography of other lands would be ploughed under and the new European seed sown therein.

Unable to converse directly with the subject races, Europe set about creating caricatures of the societies it ruled. Edward Said has labelled one of the more notorious of these instruments, 'orientalism': the impressive corpus of literature and institutions through which Europe attempted to dominate, restructure and exercise authority over the Orient. Knowledge now reflected political interests. Once racial myths were invented, it seemed far easier to interact with the myths themselves rather than with the people those myths had come to represent.

The quincentennial celebrations will focus on the 12th of October, 1992. However, Europe celebrated its conquest over the rest of human-kind throughout the period of dominance. It did so through its literature and poetry, its recorded history, its Parliamentary debates, its sciences, both natural and social, and through the tyrannical propagation of its medical knowledge. It is this political literature that *Decolonizing*

History excoriated, whether it took the form of sociology, history, anthropology, histories of science and technology, ethics. It was a language of conquest, of conquerors, clothed in the symbols of academia, of scholarly objectivity.

The calculated and pompous abuse of the rest of cultured mankind was naturally painful for the colonized: drowned in resentment, they receded further into their shells, presenting an innocuous outward facade of acquiescence to disarm and humour the colonizer. But it inevitably also raised wave upon wave of rebellious subjects, inaugurated the revolt against the white and the west, and generated a phenomenally large volume of hatred to replace the ambiance that had greeted both Christopher Columbus and Vasco da Gama when they set foot on foreign lands.

It also proved to be no less devastating for Europe as well. Ashis Nandy, J.P.S. Uberoi and others who have been engaged in a wide-ranging Europology, have demonstrated that the distortions under which an over-confident, oppressive Europe began to operate boomeranged to have a correspondingly damaging effect on the mind of Europe itself, an illness from which it is yet to recover. The decision to treat those outside the pale of Europe as aliens or as opponents prevented Europe not only from identifying with them but seeing them as an integral part of its own experience.

Predictably, today, the grand social engineering project to re-make Asia or South America in the image of Europe lies in tatters. The entire imported conceptual and behavioral edifice of bourgeois civil society and its institutions is on the verge of collapse: we have come to discover it jars with our own traditions, perceptions and habits. Today, post-independence, we still use the inherited (not chosen) outward structure, but the soul has gone out of it. Perhaps it never had any soul. It had been imposed to function as an instrument of power and domination.

The reasons for this nearly wholesale rejection of European civilization have been analyzed and studied in numerous essays but almost the most compelling of them is that people are born as cultural artefacts with distinct preferences, a major theme in *Decolonizing History*. People *qua* subjects often accept things that they would never do as free human beings. Once the overlord had been forced to sail out, taking his baggage and furniture with him, and the clouds of oppression lifted, people, like compressed molecules, would expand to new, more comfortable positions. A few Asian leaders (Jawaharlal Nehru, Tunku Abdul Rehman) might attempt to hold the flag of Europe flying aloft for a while. But theirs was a transitional generation. The one that followed it would have precious little reason to be loyal to a distant civilization, and less disagreeable to dismissing the encounter with it as an aberration or a nightmare.

It stands to reason that, post-independence, the suppressed civilizational urges and preoccupations of the South, its Ramayanas and Mahabharatas, its instinctual desire to survive, were bound to assert themselves and regain position centre-stage. As with a palimpsest, the earlier civilizational imprint would be apparent just beneath the surface. Unlike a palimpsest, the imprint would be alive and active.

Putting it in another way, once 'Europe' lost its political control over the rest of humankind, its antipode, the 'non-European' world, simultaneously ceased to exist as a counter-point. It became difficult to see India or Asia in terms of being a nondescript 'non-European' mass any longer. India is not non-West, it is India, thundered Nandy in *The Intimate Enemy*, not without reason.

Thus, just as in 1492, we discover that countries like India and China have once again returned to their own priorities and values. China and India between themselves number close to two billion people. Extremely large societies, by the time they are through with their own immediate tasks they have precious little time for what goes on outside their borders. India and significant regions in the South have also been political centres in their own right. They see Europe and the rest of the world as mere adjuncts to their own cultural projects. In fact, they are unable to see Europe as essential to their scheme of things. Often they do not see Europe at all, except sometimes on an advertising billboard which may exalt European trivia like perfumes or underwear.

Now this is not because countries like India or Malaysia or Singapore are growing industrial powers, and not because countries like India have become regional powers either. These phrases ("industrial/regional powers") only serve to conceal depressing fictions.

The basis of Asia's strength lies in the restoration of *homo faber*, in the renewed ability of millions of its peasants to produce their own food and survive even in the most difficult and adverse of circumstances, and to successfully manage and conceal their inner world from the prying hands of those who are still trying desperately to control them either through the propagation of scientific temper, development or progress.

Is the new self-preoccupation of countries like India and China then just a new parochialism evoked to fight and replace the older Eurocentrism now disgraced? Is the continuing atrocious behaviour of Europe in planning festivities around the idea of conquest a justification for rejecting the quest for human unity?

In his introduction, Rajni Kothari readily acknowledges the inordinate parochialism of the West, of its science and its objectivity. However, he goes on to reject the notion that cultural relativism can be a satisfactory alternative to hegemonistic universalism. He insists that if the West and its institutions are not to be the basis for world unity, the solution is not to give up the quest for unity altogether but to find a new basis for such unity instead.

The moot point is whether the quest for such unity or its basis is indeed a felt need of divergent cultures. In *Decolonizing History*, I fought for maintaining the radical diversity of cultures on the ground that man is irrevocably ethnic, and the idea of universality is meaningless and empty.

Fortunately at this moment, unlike as in 1492, Europe is no longer looking outward, but within. The orientation inwards is necessitated by attempts to resolve and iron out the trying and difficult problems associated with becoming a new political entity, the United States of Europe. The dismantling of communist Europe has also raised a fresh crop of problems that need urgent attention. All this simply means that Europe may not have much time to devote to the preoccupations that moved it to other parts of the world over the last five hundred years.

In that sense, the organization of the nations of Europe into a United States of a sort, as a distinct formal political entity, now dawning on us, may sharpen rather than shorten the felt distance between Europe and the societies of the South. For now, unlike as in 1492, Europe seeks to create a new self-identity without recourse to a counterpoint.

One clear sign of this, besides the quincentennial celebrations, is the attempt to make Queen Isabella (the sovereign who sponsored the Columbus journey), a saint of the Holy Roman Catholic Church and therefore a special patron of Europe. The Church was an active ingredient of the 500 years of violence and conquest. How very revealing that when more than half the globe is screaming for a fresh commitment to equality and justice, Christian Europe in blatant disregard should wish to canonize Isabella!

The question to ask here is whether Europe is behaving out of necessity or out of choice. Freedom is the recognition of necessity. It is only in this charitable sense that one accepts the amalgamation of Europe. The question remains: is Europe free from the demonology that drove those who led it in the last 500 years?

The answer to that question is perhaps already settled. By technology. And the form of social organization the technology has demanded and got.

Part of my dread of Europe came with my everyday experience of homogeneity in Europe - the dreadfulness of Europe - the pointless drive for productivity, efficiency, economy, achievement, which I found culturally abhorrent and therefore politically repulsive. The urge to homogenize the globe, eradicate tradition, flatten cultural diversity, manufacture homogenous individuals first sparked in 1492 was still alive and kicking. The Western economies had nearly achieved this quest with regard to their own citizens. In our part of the world, this quest has become effectively mired in the mud despite stimulation from the World Bank and other influential capitalists. And steadily drowning in a battle that may take another 500 years.

I have documented a history of anti-development wars in my next book, a sequel to *Decolonizing History*, titled, *Science, Development and Violence* (also being published in this year by Oxford University Press).

True freedom can only come when we are permanently liberated from any further imposition of Europe's image of us on us. Not that Europe still does not glitter. Or that America does not beckon as a promised land. But ordinary people within the more ancient civilizations seem to have come to the conclusion that the kind of organized life and living modern technology demands is ill-suited to their own view on life. The reduction of human beings to abject consumers on the one hand, and 'productive' and 'efficient' cogs in a mindless megamachine on the other, both seem somehow an affront to human history as we have known it.

But that is only part of the problem. The principal one still remains that for the past five hundred years, little that Europe did indicated that a superior civilization was at work, one that had to be imitated or incorporated with zeal, love and admiration, not fear.

Maybe in a distant future, after we have found ourselves, and Europe for its part has come to terms with the inevitability of accepting the rest of humankind as equals, we may dream of a new voyage of discovery of mutual attractions that will allow a more harmonious relationship than has existed in the past 500 years. Till that comes to pass, we who refuse the invitation to be Europeans must reject the unity of man as well.

Parra 403 510
Goa, India.

Claude Alvares

Notes and Annotated Bibliography

AN APOLOGIA AS PRELUDE

1. Barraclough G, *An Introduction to Contemporary History* (1964/1974).
2. I am aware of the equally objective world history of William McNeill, *The Rise of the West: a History of the Human Community* (1962). See also McNeill's piece, *World History in the Schools,* where he discusses these issues. The essay is included in Ballard M (ed) *New Movements in the Study and Teaching of History* (1970). See also, Wertheim W.F., *Asian History and the Western Historian,* in *Bijdragen Taal-Land-en Volkenkunde,* Vol 119 (1963) pp 150-160 for a spirited discussion.
3. Lopez, Barnes, Blum, Cameron, *Civilizations: Western and World* (1975). For a further illustration, see, *The Timetables of History: A Chronology of World Events from 5000 BC to the Present Day* (1975). Despite the "world" title, the author admits later in an introduction that the focus of the work is actually Western Europe and the Americas. The volume was prepared by Bernard Grun.
4. Romein J, *Asian Century*(1949), recognizes this point in the introduction.
5. Other civilizations including the Islamic and the Chinese have also attempted such postures. See Fitizgerald C.P., *The Chinese View of their Place in the World* (1969) for a brief and superb introduction to the Chinese position. The major difference between China and Europe seems to have consisted in the Europeans being more successful in convincing others of European superiority.
6. Hanson H.W., *A History of Art* (1970).
7. MacIntyre A, *A Short History of Ethics* (1974).
8. *Cultures and Time* (1976), published by UNESCO.

INTRODUCTION

1. Geertz C, *The Interpretation of Cultures* (1973) p 5.
2. Benedict R, *Patterns of Culture* (1934).
3. Gardiner P, *Herder,* in Edwards P (ed), *The Encyclopaedia of Philosophy,* Vol 3 (1972) pp 486-489.
4. Cole M, Gay J, Glick A.J., Sharp D.W., *The Cultural Context of Learning and Thinking* (An Exploration in Experimental Anthropology) (1971) contains the evidence.
5. Huizer G & Mannheim (eds), *The Politics of Anthropology* (Forth-

coming Mouton: 1977. Clark L.E. (ed), *Through African Eyes* (1971) avoids this "imperialism of categories". I am grateful to Ward Morehouse for drawing my attention to this volume.

6. Chesneaux J, *Peasant Revolts in China: 1840-1949* (1973).
7. Inglis B, *Poverty and the Industrial Revolution* (1971).
8. For a lucid introduction to the sociology of knowledge, Hamilton P, *Knowledge and Social Structure* (1974). The power of the West to decide and determine what constitutes knowledge, particularly in relation to the Southern countries, is discussed in Goonatilake S. *Afhankelijke Sociale Structuren en de Sociale Structuur van kennis afhankelijkeid, Internationale Spectator* (1976).
9. Lach D., *Asia in the Making of Europe*, 2 vols. (1965).
10. Wertheim W.F., *Asian History and the Western Historian* (1963) p 151.
11. Drucker P, *The Landmarks of Tomorrow* (1959).
12. Needham J, *Between the Four Seas* (1969).
13. Halliday J & McCormack G, *Japanese Imperialism Today* (1973); Achterhuis H, *Filosofen van de Derde Wereld* (1975).
14. Benedict R, *The Chrysanthemum and the Sword: Patterns of Japanese Culture* (1946).
15. Horowitz I.L., *The Rise and Fall of Project Camelot*, MIT Press.
16. Chomsky N, *American Power and the New Mandarins* (1969). See particularly the first chapter.
17. For a detailed analysis of the Ford Foundation and Rockefeller-dominated American movement for birth control in Southern countries, see Mass B, *The Political Economy of Population Control in Latin America* (1975), editions Latin America, Canada. The role of the World Bank and the IMF in the compulsory sterilization programmes during the Indian Emergency of 1975-76 has been analysed by Novak J, *The Role of IMF, World Bank, The Times of India,* 1, 4, 5 July 1977.
18. For an overall discussion, see Barraclough G. series of articles in the *New York Review of Books* on the new world economic order (1976).
19. Wertheim W.F., *Evolution and Revolution* (1974).
20. Bauer P, *Dissent on Development* (1972).
21. Lefeber L, *On the Paradigm for Economic Development,* in *World Development*. Vol 2. Jan 1974. p. 1.
22. Lerner D, *The Passing of Traditional Society: Modernizing the Middle East* (1958).
23. Kothari R, *State-building in the Third World: Alternative Strategies,* in *Economic and Political Weekly* (1972).
24. Frank A.G., *Latin America: Underdevelopment or Revolution* (1960). For a consummate summary and critique of all the major "development" theories concerning Latin America, see Bertholet C.J.L., *De Strijd om een Adequaat Ontwikkelings-paradigma voor de Derde Wereld* (1975). An English version is available from the author at the University of Tilburg, Holland.
25. Hesseling Pjotr, *Organizational Behaviour and Culture: The Case of the Multinational Enterprise* (1971).
26. Rudolph L & Rudolph S, *The Modernity of Tradition* (1967); see also

that fine article by Isma'il R. Al Faruqi, *Science and Traditional Values in Islamic Society*, in Morehouse W (ed), *Science and the Human Condition in India and Pakistan* (1968).
27. Heesterman J.C., *India and the Inner Conflict of Tradition*, in *Daedalus*, Winter 1973. Vol 102. pp 97-114.
28. Rudolph & Rudolph, *The Modernity of Tradition* (1967).
29. McClelland D, *The Achieving Society* (1961).
30. Myrdal G, *The Challenge of World Poverty* (1970).
31. Foster G.M., *Peasant Society and the Image of the Limited Good*, in *The American Anthropologist*. Vol 67 (1965) pp. 293 ff.
32. Banfield E.C., *The Moral Basis of Backward Society* (1958).
33. Erasmus C., *Community Development and the Encogido Syndrome*, in *Human Organization*, Vol 27 (1968) pp 70 ff.
34. Huizer G, *Rural Extension and Peasant Motivation in Latin America and the Caribbean* (1973). See also his *Revolutionary Potential of Peasants in Latin America* (1972) and, *Peasant Rebellion in Latin America* (1973).
35. Wertheim W.F., *Resistance to Change – For whom?* in Evers Hans-Dieter (ed), *Modernization in South-East Asia* (1973). See also, Wertheim's *East-West Parallels: Sociological Approaches to Modern Asia* (1964).
36. Makhijani A, *Energy Policy for the Rural Third World* (1976) p 5.
37. The concept of a survival algorithm may be found discussed in Streeten P & Lipton M, *The Crisis of Indian Planning* (1968).
38. Heilbroner R, *The Future as History* (1960).
39. Heilbroner R, *op. cit.* For another example, see Marglin S, *Irrigation Planning and Research in India and West Pakistan*, in Morehouse W, (ed) *op cit:* pp 189-203.

CHAPTER ONE

1. Bronowski J, *The Ascent of Man:* pp 115-116 (1973)
2. Douglas M, Introduction to Dumont L, *Homo Hierarchus* (1972) p. 13.
3. Mumford L, *Technics and Human Development* (1967) p 5.
4. Published in 1974, by Methuen.
5. Review of Thorpe's book, *TLS*. 28 February 1975.
6. Mumford L, *op. cit.*, p. 6.
7. Geertz C, *The Interpretation of Cultures/p 61.*
8. Huizinga J, *Homo Ludens* (1937-1950) Foreword.
9. Huizinga J, *op. cit.*, p. 173.
10. Kerr W, *The Decline of Pleasure* (1962).
11. Cassirer E, *Essay on Man* (1944). For a very significant contribution on the relationship between language and cosmology, see Whorf B.L., *Language, Thought and Reality* (1956).
12. Singer P, *Animal Liberation* (1975).
13. Kwee Swan-Liat, *Mens en Gereedschap* in *Is Vakmanschap*

Meesterschap (Technische Hogeschool, Eindhoven). 1975 pp 3.1-3.6. See also his *De Mens Tussen Mythe en Machine* (1974).
14. Von Frisch K, *Animal Architecture* (1976). Also interesting and not irrelevant is Paturi F, *Nature, Mother of Invention* (1975).
15. Geertz C, *op. cit.*, p. 49.
16. Gasset Ortega y, *Man the Technician* in Mitcham & Mackay (eds), *Philosophy and Technology* (1972) p 296.
17. Geertz C. *op.cit.*, pp 49-51.
18. Kwee Swan-Liat, *Mens en Gereedschap. Ibid.*
19. Lewis & Towers, *Naked Ape or Homo Sapiens?* (1972) gives a lucid account of the term "specialization" in the evolutionary context.
20. Durant W, *The Pleasures of Philosophy* (1953) p. 263.
21. Williams C, *Craftsmen of Necessity* (1974). See also, Wulff Hans, *The Traditional Crafts of Persia* (1966), and Rudofsky B, *Architecture without Architects* for further illustrations.
22. Spier R, *From the Hand of Man* (1970) p 12.
23. See the work of McKim Marriott, who visited an isolated village in 1969 in Uttar Pradesh after an earlier visit in 1951. Marriott reports significant and far-reaching social changes, based in part upon changing technology. A visual presentation of his findings is available at the Educational Resources Centre, New Delhi. I am grateful to Prof. Morehouse for drawing my attention to Marriott's work.
24. Meadows et al, *Limits to Growth* (1970) since superseded by *Mankind at the Turning Point* (1975) which argues for selective growth.
25. Toynbee A, *A Study of History* Vol III p. 386.
26. Kranzberg M & Davenport W.H., *Technology and Culture* (1972) p 17.
27. Gasset, *op. cit.*, p 306.
28. To be brief, White L, *Tibet, India and Malaya as Sources of Western Medieval Technology*, in *American Historical Review* (1960) 65. 515. Childe G.V., *The Oriental Background of European Science*, MQ (1938) 1. 105. The largest amount of material is to be located in Dr Joseph Needham's copious works on China. White L, *Medieval Technology and Social Change* (1962/75) and Braudel F, *Capitalism and Material Life: 1400-1800* are also useful.
29. Bacon F, quoted in Needham, SCC I, p 19.
30. Needham J, *History of Science and Technology in India and South-east Asia*, in *Nature*. 14 July 1951 p. 64.
31. Gunawardena R.A.L.M., *Hydraulic Engineering in Ancient Sri Lanka: the Cistern Sluices*, in *The Journal of the Institute for South Asian Archaeology* (1976).
32. Sivin N, in the foreword to Jeon Sang-woon's *Science and Technology in Korea* (1974) p xii.
33. Sivin N, *op. cit.*, p xv.
34. Murphy E.J., *History of African Civilization* (1972) p xviii. The volume also contains a comprehensive bibliography. Goody J, *Technology, Tradition and the State in Africa* (1971) is a slim volume, but excellent.
35. Rodney Walter, *How Europe Underdeveloped Africa* (1976) p 81.

36. Goody J, *op. cit.*, p 23.
37. Forbes R, *Studies in Ancient Technology*. Vols VI & VIII. For a contemporary picture, see Müller Jens, *Decentralized Industries and Inadequate Infrastructure* (1976), Copenhagen Institute for Development Research. I am grateful to Ward Morehouse again for sending me this paper.
38. Davidson B, *Old Africa Rediscovered* (1959). *The African Past,* by the same writer, contains documents covering a period from ancient Egypt to the present (1964). The first volume also contains an extensive bibliography.
39. Davidson B, *OAR:* p 58.
40. Davidson B, *OAR:* p 72.
41. Davidson B, *OAR:* p 73.
42. Davidson B, *OAR:* p 144.
43. Robinson, *A History of Dyed Textiles* (1969) p 77. See also, *A History of Printed Textiles,* by the same author. For a wide range of crafts, see Trowell M, *African Design* (1960). Also, *CIBA Review,* No 21 on *Weaving and Dyeing in North Africa,* and Oliver Paul, *Shelter in Africa* (1971).
44. Davidson B, *The African Personality* pp 535-40 in Legum C, (ed). *Africa Handbook.* His *Which Way Africa*? (1967) also discusses similar themes. Cesaire Aime's *Discourse on Colonialism* (1972) is as indispensable as it is lively. See also the work edited by Leon Clark.
45. For general works, see Coe M, *The Maya* (1966), which has a decent bibliography. Brundage B.C., *Empire of the Inca* (1969) contains philosophical anthropology, with an introduction by Toynbee, based on extensive original sources. For languages, see the work of Benjamin L. Whorf.
46. D'Harcourt R, *Textiles of Ancient Peru and their Techniques* (1962); see also. *CIBA Review:* No 70, Textile Art in Ancient Mexico; No 136 on Peruvian Textile Techniques.
47. Braudel F, *op. cit.,* pp 25-28.
48. Mumford L, *The City in History* (1961) p 226.
49. Clissold S, *Latin America: a Cultural Outline* (1965) p 22.
50. Clissold S, *op. cit.,* p 19.
51. Needham J, *Within the Four Seas* (1969) pp 21-24.

CHAPTER TWO

The literature spread out among the notes below is a small sample, not even representative or organized. While a great deal of the Sanskrit, Persian, Arabic, and Tamil sources have been translated, a lot more remains to be done. A detailed, but by no means comprehensive bibliography of source material in science and technology, of books written in Sanskrit, Arabic, and Persian languages has just been completed under the direction of A. Rahman: see his 1977 volume, *Triveni,* for details. According to Rahman, it gives a fair idea of the extent and range of scientific and technological activity in India during the period.

Accounts of foreign travellers in India and Asia during the period 1500-1800, necessary for reconstructing the history of the period, are listed comprehensively in the volumes of Lach D, *Asia in the Making of Europe,* two volumes, three books so far.

O.P. Jaggi's *History of Science and Technology in India,* in five volumes, is shabbily done and the title exaggerates. Three of the volumes are on Indian medicine and the other two are restricted to science and technology in the earliest periods of Indian history. More useful, and revelatory, are the volumes of M.A. Alvi and A. Rahman, *Fathullah Shirazi – A sixteenth Century Indian Scientist* (1968) and *Jahangir – The Naturalist* (1968). Rahman's piece, *Sixteenth- And Seventeenth-Century Science in India and Some Problems of Comparative Studies,* in Teich and Young (eds), *Essays in Honour of Joseph Needham: Changing Perspectives in the History of Science* (1973) contains a good introduction to the period under review. Rahman's *Trimurti: Science, Technology and Society* has a chapter on traditional Indian science and Technology. There is a further chapter by Rahman on a similar theme in Morehouse (ed), *Science and the Human Condition in India and Pakistan.*

The best work todate on the subject is Dharampal's *Indian Science and Technology in the Eighteenth Century,* with an introduction by the author. Dharampal's work should be read keeping in mind the science historian, David Pingree's critique in *The Journal of Asian Studies.* Vol XXXII, No 1 Nov. 1972 pp 178-179. Finally, Needham's work on China provides innumerable references to Indian science and technology.

The textile industry has been well researched due to a band of devoted scholars centred round the Calico Museum of Textiles in Ahmedabad, which also publishes an irregular but first-rate journal on textile history. Two issues of the *CIBA Review* are also devoted to textiles in India. The Calico Museum has also reprinted *A Select Bibliography of Indian Textiles* (1975).

1. Müller M, *A History of Sanskrit Literature.* p 3.
2. Kapp W, *Hindu Culture, Economic Development and Economic Planning in India* (1963); Mishra V, *Hinduism and Economic Growth* (1962); Myrdal G, *Asian Drama* (1968); Maddison A, *Class Structure and Economic Growth: India and Pakistan since the Moghuls* (1971); Spengler J.J., *Indian Economic Thought: a Preface to its History* (1971) goes to the extent of discussing India's "pain economy".
3. Dasgupta S, *Hindu Ethos and the Challenge of Change* (1972).
4. Watts A, *Nature, Man and Woman* (1958).
5. Chandra P, *Ideological Discord in Ancient India,* in *Quest* 99 pp 1-10.
6. Weber M, *The Religion of India* (1958) p 4.
7. Armitage W, *The Rise of the Technocrats.* pp 3-9.
8. Dutt R.C., *The Economic History of India* Vol I (1973) Reprint editions. pp 190-191.
9. Dutt R.C., *op. cit.,* p 191.
10. Dharampal, *op. cit.*
11. Walker, *Indian Agriculture,* in Dharampal, *op. cit.,* p 180-181.
12. Walker, *op. cit.,* pp 181-182.

13. Walker, *op. cit.*, p 184.
14. Walker, *op. cit.*, pp 186-187.
15. Walker, *op. cit.*, p 192.
16. Walker, *op. cit.*, pp 198-199.
17. Dutt R.C., *op. cit., p 191.*
18. Fazl A, in Sarkar J.N., *Studies in Economic Life in Mughal India* (1975) p 295.
19. Lach D, *op. cit.* Large chapters on the intricacies of the spice trade.
20. Needham & Huang Ray, *The Nature of Chinese Society: A Technical Interpretation* (1975).
21. Majumdar, Raychaudhuri, Datta, *An Advanced History of India.* 3rd Ed. (1967) p 561.
22. Walker, *op. cit.* p 194.
23. Hussain G, in Raghuvanshi V.P.S. *Indian Society in the Eighteenth Century* (1969) p 315.
24. Walker, *op. cit.*, pp 194-196
25. Dubois, in Raghuvanshi, *op. cit.*, p 319.
26. Gadgil, *The Industrial Evolution of India* (1969).
27. Irwin & Schwartz, *Studies in Indo-European Textile History* (1966) p 8.
28. De Guyon, in Raghuvanshi, *op. cit.*, p 326.
29. Irwin & Schwartz, *op. cit.*, p 15.
30. Roques G, *La Manière de negocier dans les Indes Orientales. Calico Museum Journal.*
31. Irwin & Schwartz, op. cit., p 18.
32. Singer C et al., *A History of Technology* Vol II. p 774.
33. Robinson S, *A History of Printed Textiles.* p 60.
34. Robinson S, *op. cit.*, p 62.
35. Robinson S, *op. cit.*, p 118.
36. Coeurdoux, in Irwin & Schwartz, *op. cit.*, p 104.
37. Irwin & Schwartz, *op. cit.*, p 113.
38. Robinson S, *op. cit.*, pp 112-113.
39. Roques G, *op. cit.*
40. Irwin & Schwartz, *op. cit.*, p 119.
41. Robinson S, *op. cit.*, p 15.
42. Coon C.S. *The History of Man* (1954) pp 328-329.
43. Stodart, in Heyne, *Tracts on India.* p 363.
44. Stodart, *op. cit.*, p 364.
45. Dharampal, *op. cit.*, p. LXIX. ŋ. 36.
46. Dharampal, *op. cit.*, p. LI.
47. Dharampal, *op. cit.*, p LX.
48. Filliozat J, *The Classical Doctrine of Indian Medicine* (1949); Zimmer H, *Hindu Medicine.*
49. Dharampal, *op. cit.*, p XLIII.
50. Dharampal, *op. cit.*, p 268.
51. Dharampal, *op. cit.*, pp 155-156.
52. Dharampal, *op. cit.*, p 153.
53. The discussion that follows is mainly taken from the excellent paper by Almast S.C., *History and Evolution of the Indian Method of*

Rhinoplasty, a paper delivered and included in the Proceedings of the Fourth International Congress on Plastic Surgery 1967. Excerpta Medica, Amsterdam, 1969.
54. Almast S.C., *op. cit.*, p 23.
55. Almast S.C., *op. cit.*, p 21.
56. Almast S.C., *op. cit.*, pp 23-24.
57. Needham J, SSC (1965) Vol IV. Part 2. Also, note 37, chapter one.
58. Bernstein H, *Steamboats on the Ganges: An Exploration in the History of India's Modernization Through Science and Technology* (1960).
59. Dharampal, *op. cit.*, p LVII.
60. Needham J, *Within the Four Seas* (1969) pp 182-183.
61. See Filliozat's preface to his book on Indian Medicine for a good introduction. Most of the material on Indian science is handled by the National Institute of Sciences in India, New Delhi.
62. Robinson S, *op. cit.*, p 111. See also, *CIBA Review,* no 2. *India, its Dyers and its Colour Symbolism.*
63. Lannoy R, *The Speaking Tree* (1971/74) pp 46-47.
64. Northrop F.S.C., *The Meeting of East and West* (1949).
65. Lannoy R, *op. cit.*, p 32.
66. Lannoy R, *op. cit.*, pp 43-44. See also, Pacey A, *The Maze of Ingenuity* (1974) p 287 ff.
67. Lannoy R, *op. cit.*, p 277.
68. Lannoy R, *op. cit.*, p 275.

CHAPTER THREE

1. Kiernan V.G., *The Lords of Human Kind* (1972); Sansom G, *The Western World and Japan* (1949/73); Panikkar K.M., *Asia and Western Dominance* (1953); Lach D, *Asia in the Making of Europe* (1965/1970).
2. See, Castro Josue de, *Compass of China* in *Comprendre.* 19. p 33.
3. Wertheim W.F., *The Better Earth* in *Comprendre.* 19. p 108.
4. Toulmin & Goodfield, *The Architecture of Matter* (1962) pp 25-26.
5. Toulmin & Goodfield, *op. cit.*, p 42.
6. Price Derek J de Solla, *Is technology historically independent of science?* in *Technology and Culture,* Fall, 1965, Walt Rostow's recent book, *How It All Began* (1975) betrays a fundamental misunderstanding of these points, which weakens the central argument of his thesis.
7. Sivin N, in Nakayama & Sivin (eds), *Chinese Science, Explorations of an Ancient Tradition* (1973) p xiii.
8. Nasr S.H. *Islamic Science: An Illustrated Study* (1976). The volume has a good bibliography. See also his *Science and Civilization in Islam,* based on literary sources.
9. Nasr S.H., *op. cit.*, preface p xiii.
10. Sivin N, in Nakayama & Sivin, *op. cit.*, p xviii.
11. Sivin N, in Nakayama & Sivin, *op. cit.*, p xviii.
12. Sivin N, in Nakayama & Sivin, *op. cit.*, p xvii.

13. Nakayama S, in Nakayama & Sivin, *op. cit.*, p 36.
14. Needham J, *On Science and Social Change* in *The Grand Titration: Science and Society in East and West* (1969) p 150.
15. Needham J, *op. cit.*, p 40.
16. Graham A.C., *China, Europe and the Origins of Modern Science,* in Nakayama & Sivin, *op. cit.*, pp 53-54.
17. Graham A.C., *op. cit.*, pp 60-61.
18. Hooykaas R, *Religion and the Rise of Modern Science* (1973) p 162.
19. Elvin M, *The Pattern of the Chinese Past* (1973) pp 233-234.
20. Hooykaas R. *op. cit.*, p 139.
21. Yates F, *Science, Salvation and the Cabala in NYR of Books,* 27 May (1976) p 27.
22. Graham A.C., *op. cit.*, p 59.
23. Graham A.C. *op. cit.*, p 51.
24. Kwee Swan-Liat, *Mens en Natuur in het Chinese Denken* in *Wijsgerig Perspectief op Maatschappij en Wetenschap* (1973-4) No. 6. p 345.
25. Kwee Swan-Liat, *op. cit.*, p 344.
26. Kwee Swan-Liat, *op. cit.*, pp 352-353.
27. De Bary W.T., *Chinese Despotism and the Confucian Ideal: A Seventeenth Century View* in Fairbank J (ed), *Chinese Thought and Institutions* (1957). Much of the discussion on this issue comes from de Bary's essay on Huang Tsung-hsi, but the overall interpretation is my own.
28. Quoted in de Bary, *op. cit.*, p 168.
29. Quoted in de Bary, *op. cit.*, p 169.
30. De Bary W.T., *op. cit.*, p 170.
31. De Bary W.T., *op. cit.*, p 172.
32. Needham J, *Within the Four Seas.* p 91.
33. Nakayama S, in Nakayama & Sivin, *op. cit.*, p 35.
34. Hommel A. W., *American Historical Review* (1955) pp 610-612.
35. Fung Yu-lan, *The International Journal of Ethics* (1922) pp 237-263.
36. Fung Yu-lan, *op. cit.*
37. Fung Yu-lan, *op. cit.*
38. Sivin N, in Nakayama & Sivin, *op. cit.*, pp xxix-xxx.
39. Nasr S.H., *op. cit.*, pp 15-17.
40. Needham J, *Within the Four Seas.* p 82.
41. Needham & Huang Ray, *The Nature of Chinese Society. East West Journal* (1974-75).
42. Needham J, *SCC.* Vol I. pp 51-54.
43. Sansom G, *op. cit.*, p 103; Panikkar K.M., *op. cit.*, pp 120-121.
44. Needham J, *Hand and Brain in China* (1971). I am indebted to Ward Morehouse for introducing me to two earlier works, by Derk Bodde, produced by the American Council on Education, *China's Gifts to the West* (1942-71) and *Chinese Ideas in the West* (1948-65).
45. Flessel K, *Der Huang-Ho und die Historische Hydrotechnik in China.* Tubingen (1974) pp 210-211.
46. Pacey A, *The Maze of Ingenuity* (1974) pp 289-290.
47. Dietrich C, *Cotton Culture and Manufacture in Early Ch'ing China* in

Willmott W.E. (ed), *Economic Organization in Chinese Society* (1972) p 109. I am indebted to Dietrich and to Mark Elvin for most of the material used in this discussion.

48. See Elvin M, *The Pattern of the Chinese Past* (1973) p 212.
49. See Dietrich C, *op. cit.*, p 111.
50. Dietrich C, *op. cit.*, p 126.
51. Elvin M, *op. cit.*, p 253.
52. Elvin M, *op. cit.*, pp 179-180.
53. Sun E-Tu Zen, *Sericulture and Silk Textile Production in Ch'ing China* in Willmott, *op. cit.*
54. Elvin M, *op. cit.*, p 284.
55. Dean C.G. *Science and Technology in the Development of Modern China: An Annotated Bibliography* (1974). For traditional Chinese science and technology, see pp 201-225.
56. Fitzgerald C.P., *The Birth of Communist China*. p 35.
57. Needham & Huang, *op. cit.*, 381.
58. Ho Ping-ti, *Studies on the Population of China: 1368-1953* (1959).
59. Perkins D.H., *Agricultural Development in China: 1368-1968* (1969). I have used Perkins' modifications of Ping-ti's figures on the population of China.
60. Elvin M, *op. cit.* See also his earlier essay in Willmott (ed), *op. cit.*, pp 137-172.
61. Braudel F, *Capitalism and Material Life*. p 10.
62. Boserup E, *The Conditions of Agricultural Growth* (1965).
63. Perkins D.H., *op. cit.*, p 6.
64. Perkins D.H., *op. cit.*, p 6.
65. White L, *Medieval Technology and Social Change* (1962).
66. Wilkinson R, *Poverty and Progress* (1973).
67. Wertheim W.F., *The Better Earth. op. cit.*, p 103.
68. Braudel F, *op. cit.*, p 248.
69. Elvin M, *op. cit.*, p 314.

CHAPTER FOUR

1. See the following important, key works:
   Mantoux P, *The Industrial Revolution in the Eighteenth Century* (1961); Ashton T.S., *The Industrial Revolution: 1760-1830* (1975); Landes D.S., *The Unbound Prometheus: Technological Change and Industrial Development in Western Europe from 1750 to the Present* (1969); Giedion S, *Mechanization Takes Command* (1948); Hobsbawm E.J., *Industry and Empire: From 1750 to the Present Day* (1968/1975); Derry T.K. & Blakeway M.G., *The Making of Pre-Industrial Britain* (1969/1973); Wilkinson R, *Poverty and Progress* (1973); Inglis B, *Poverty and the Industrial Revolution* (1971); Hodgen M.T., *Change and History* (1952).
2. Ashton T.S., *op. cit.*, p 2.
3. Forbes R.J., *The Conquest of Nature* (1968) p 43.
4. Nef J.U., *The Progress of Technology and the Growth of Large-Scale*

Industry in Great Britain: 1540-1640 in Carus-Wilson (ed), *Essays in Economic History*. Vol I. 1954.

5. Wertheim W.F., *Evolution and Revolution* (1974).
6. Cipolla C.M., *The Economic History of World Population*. 6th Ed. (1974). White L, *Medieval Technology and Social Change.*
7. Boserup E, *The Conditions of Agricultural Growth: The Economics of Agrarian Change under Population Pressure* (1965); see also, Joan Robinson's use of this in *Freedom and Necessity: An Introduction to the Study of Society* (1970); see for further confirmation, the new volume by Mark Nathan Cohen, *The Food Crisis in Prehistory: Overpopulation and the Origins of Agriculture* (1977), which argues that the nearly simultaneous worldwide emergence of agriculture can be explained by a single phenomenon: the growth of world population beyond the saturation point for hunting economies.
8. Sahlins M, *Stone-Age Economics* (1974) chapter one. A more recent work is Clarke R & Hindley G, *The Challenge of the Primitives* (1975).
9. See Robinson J, *op. cit.*
10. Boserup E, *op. cit.*, p 38. For allied evidence, see North D.C. & Thomas R.P., *The Rise of the Western World* (1973).
11. Postan M.M., *The Medieval Economy and Society* (1972/75).
12. Fischer F.J., *Inflation and Influenza in Tudor England, Economic History Review* XVIII (1965).
13. Elvin M, *The Pattern of the Chinese Past*. pp 120-121.
14. See Derry & Blakeway, *op. cit.*, p 67.
15. More T, *Utopia* (Turner translation). p 46.
16. See Heilbroner, *The Worldly Philosophers* (1953/72) pp 29-30
17. Ashton T.S., *op. cit.*, pp 20-21
18. Landes D, *op. cit.*, p 115.
19. Inglis B, *op. cit.*, p 88.
20. White L, *op. cit.*, p 39.
21. White L, *op. cit.*, p 40.
22. See Inglis B, *op. cit.*, p 66.
23. See Inglis B, *op. cit.*, p 69.
24. Childe G, *Man Makes Himself* (1951) p 18.
25. Omvedt G, *The Political Economy of Starvation* (1975) p 15.
26. North & Thomas, *op. cit.*
27. Roberts Glyn, *Questioning Development* has a fine example or parody on the experts in the fishing village. But see, Wilkinson R, *op. cit.*, pp 84-85.
28. Wilkinson R, *op. cit.*, p 112.
29. Braudel F, *op. cit.*, p 266.
30. Armitage W.H.G., *A Social History of Engineering* (1961).
31. Hobsbawm E.J., *op. cit.*, pp 46-47.
32. Derry & Blakeway, *op. cit.*, p 238.
33. See Landes D, *op. cit.*, for a brief but excellent survey of the technical aspects of this issue. 88 ff.
34. Salaman R.N., *The History and Social Influence of the Potato* (1949).
35. See Inglis B, *op. cit.*, for a comprehensive account of the condition of

Chimney-sweeps and for other forms of child labour.

36. See Revelle R, *Energy Use in Rural India, Science*. Vol 192. 4 June 1976, for a discussion of this in a contemporary situation.
37. Landes D, *op. cit.*, p 96.
38. Landes D, *op. cit.*, p 126.
39. Landes D. *op. cit.*, p 95.
40. For some excellent visual material of intricate but increasingly cumbersome machinery in the Limburg mines in Holland, see Raedts C.E.P.M., *De Opkomst, de Ontwikkeling en de Neergang van de Steenkolenmijnbouw in Limburg* (1974).
41. Wilkinson R, *op. cit.*, pp 123-124.
42. Wilkinson R, *op. cit.*, pp 124-125.
43. Landes D, *op. cit.*, 80 ff.
44. Derry & Blakeway, *op. cit.*, p 95.
45. Engels F, *The Condition of the Working Class in England* (1958) pp 78 79.
46. Pacey A, *op. cit.*
47. See Inglis B, *op. cit.*, p 367. See also, Hobsbawm E.J., *Labouring Men: Studies in the History of Labour* (1964/68).
48. See Inglis, B, op. cit., pp 369-70.
49. Inglis B, *op. cit.*, p 369.
50. See Inglis B, *op. cit.*, p 371.
51. Hobsbawm E.J. *Industry and Empire*. p 94.
52. Hobsbawm E.J. *op. cit.*, p 80.
53. Hobsbawm E.J. *op. cit.*, p 91.
54. See Rosenberg B & White D.M., *Mass Culture* (1957). Also, Gasset Ortega y, *The Revolt of the Masses* (1930/1972).
55. Wilkinson R, *op. cit.*, p 185.

CHAPTER FIVE

1. Lach D, *Asia in the Making of Europe*. Vol I. (1965).
2. Sansom G, *The Western World and Japan*. pp v-vi.
3. Panikkar K.M., *Asia and Western Dominance*. p 66.
4. See Teng & Fairbank, *China's Response to the West* (1971) p 19.
5. See Panikkar K.M., *op. cit.*, p 94.
6. See Panikkar K.M., *op. cit.*, p 74.
7. Sansom G, *op. cit.*, p 145.
8. Irwin & Schwartz, *Studies in Indo-European Textile History* (1966) p 10. A great deal of the information presented here on this issue comes from this excellent volume.
9. See Sansom G, *op. cit.*, p 63 n.
10. Panikkar K.M., *op. cit.*, p 66.
11. Simkin C.G.F., *The Traditional Trade of Asia* (1968) pp 163-164. See also the more detailed and scholarly dissertation of Meilink-Roelofsz M.A.P., *Asian Trade and European Influence in the Indonesian Archipelago between 1500 and about 1630* (1962).

12. Sansom G, *op. cit.,* p 103.
13. Harrison J, *The Era of the Companies* in, *Europe and the Indies* (1970) BBC Publication. p 4.
14. Sansom G, *op. cit.,* pp 148-149.
15. Spear P, *op. cit.,* p 70.
16. Lannoy R, *The Speaking Tree.* p 16.
17. Gopal R, *British Rule in India: An Assessment* (1963) p 10.
18. Gopal R, *op. cit.,* Chapter one contains a good description, but see also, Dutt R.C., *The Economic History of India.* Vol I. (1970) Reprint.
19. See Panikkar K.M., *op. cit.,* p 101. An English writer, Bolt, wrote that "instances have been known of their cutting off their thumbs, to prevent their being forced to wind silk". (Gopal R, *op. cit.,* p 7). Jens Müller, reporting from Africa, notes that the colonial administration there (also British: Tanzania) prohibited iron smiths from producing iron, on the pain of having their hands amputated. Muller claims to have interviewed some of these handless blacksmiths. So the story of the weavers cutting off their thumbs is not as far-fetched as it is made out to be.
20. For literature on the "Drain" theory, see Naoroji D, *Poverty of India* (1888); Dantwala M.L., *Poverty in India: Then and Now 1870-1970* (1973); Dutt R.C., *The Economic History of India,* both volumes.
21. Maddison A, *Historical Origins of Indian Poverty, Banca Nazionale Del Lavoro Quarterly Review* (1970) pp 50-51.
22. Dutt R.C., *op. cit.,* Vol I. pp 180-181.
23. Dutt R.C., *op. cit.*, Vol I. pp 208-209.
24. Dutt R.C., *op. cit.*, Vol I. p. 179.
25. Landes D, *The Unbound Prometheus*, p 86.
26. Deane P, *The First Industrial Revolution* (1965) p 89.
27. Landes D, *op. cit.,* p 86.
28. See Dutt R.C. *op. cit.,* Vol II. pp 85-86.
29. Harnetty P, *Imperialism and Free Trade: Lancashire and India in the Mid-Nineteenth Century* (1972) p 51.
30. Harnetty P, *op. cit.,* p 6.
31. Harnetty P, *op. cit.,* p 29.
32. Harnetty P, *op. cit.,* p 34.
33. Harnetty P, *op. cit.,* p 32.
34. Harnetty P, *op. cit.,* pp 66-67.
35. See Muller J, for African evidence: *Decentralized Industries and Inadequate Infrastructure* (1976) pp 32-33.
36. Dutt R.C., *op. cit.,* Vol I. pp 35-36.
37. Woodruff W, *Impact of Western Man: A Study of Europe's Role in the World Economy 1750-1960* (1966) p 9.
38. Darling M.L., *The Punjab Peasant in Prosperity and Debt* (1947) p 178. It is possible the rise of the moneylender was also related to the influx of people into the area. In any case, the only people to appreciate the presence of the moneylender seem to be the debtors. It should not be forgotten that the moneylender also took risks. My point is he might have kept production at below *normal* levels.
39. Mamdani M, *The Myth of Population Control: Family, Caste and Class*

in an Indian Village (1972). See particularly the superb chapter on *Technology and Social Structure.*

40. Mamdani M, *op. cit.*
41. See Mamdani M, *op. cit.*
42. Woodruff W, *op. cit.*, p 126.
43. Curwen Charles, *The Imperial Pushers. TLS,* 11 June, 1976, which reviews three new volumes on the Opium Trade.
44. Gopal R, *op. cit.*, pp 98-99.
45. Gopal R, *op. cit., p 86.*
46. Gopal R, *op. cit.*, pp 13-14.
47. Erikson E.H., *Gandhi's Truth* (1970) pp 443-448.
48. Woodruff W, *op. cit.*, pp 127-128.
49. Panikkar K.M., *op. cit.*, p 133.
50. Simkin C.G.F., *op. cit.*, p 275.
51. Panikkar K.M., *op. cit.*, p 182.
52. Panikkar K.M., *op. cit.*, p 182.
53. Panikkar K.M., *op. cit.*, p 183.
54. Elvin M, *op. cit.*, pp 313-314.
55. Gittings J, *A Chinese View of China* (1973) p 69.
56. Simkin C.G.F., *op. cit.*, p 281.
57. Morris D.M. et al, *Indian Economy in the Nineteenth Century: A Symposium* (1969). Morris proposes, *Toward a Reinterpretation of Nineteenth Century Indian Economic History.* Others attempt to dispose.
58. Morris D.M., *op. cit.*, pp 166-167.
59. Pearson M.N., *Merchants and Rulers in Gujarat: The Response to the Portuguese in the Sixteenth Century* (1977) seems to be a step in the direction of this interpretation.
60. Raychaudhuri T, *European Commercial Activity and the Organization of India's Commerce and Industrial Production 1500-1750* in Ganguli B.N. (ed), *Readings in Indian Economic History* (1964) pp 75-76.
61. Pearse A.S., *The Cotton Industry of India* (1930) p 25.
62. Morris D.M., *op. cit.*, 160 ff.
63. Sovani V.V. *British Impact on India* (two parts), in the *Journal of World History* Vol. VI. p 95.
64. Feuerwerker A, *The Chinese Economy: 1870-1911.* p 17.
65. Skinner G.W., *Marketing and Social Structure in Rural China* in the *Journal of Asian Studies* (1965) 24. Three parts: 1964/65/65.
66. Murphey R, *The Treaty Ports and China's Modernization: What went wrong?* (1970) p. 27.
67. Murphey R, *op. cit.*, p 26.
68. Murphey R, *op. cit.*, p 14.
69. Murphey R, *op. cit.*, p 41.
70. Murphey R. *op. cit.*, p 20.
71. Murphey R, *op. cit.*, p 20.
72. Murphey R, *op. cit.*, p 16.
73. Murphey R, *op. cit.*, p 32.

74. Murphey R, *op. cit.*, p 32.
75. Elvin M, *Skills and Resources in late Traditional China* in Perkins D.H. (ed), *China's Modern Economy in Historical Perspective* (1975).
76. Murphey R, *op. cit.*, p 18.
77. Feuerwerker A, *The Chinese Economy: 1870-1911.* p 17.
78. Feuerwerker A, *op. cit.*, 1912-1949. p 12.
79. Feuerwerker A, *op. cit.*, 1912-1949. p 13.
80. Dharampal's private notes, resting in the Indology Institute, University of Leiden, Holland.

CHAPTER SIX

For a comprehensive list of materials concerning the technological experience of contemporary China, see C.G. Dean's invaluable, *Science and Technology in the Development of Modern China: An Annotated Bibliography* (1974). The following works may also be deemed essential:

Feuerwerker A, *China's Early Industrialization: Sheng Hsuan-huai and Mandarin Enterprise* (1958); Feuerwerker A, *The Chinese Economy ca. 1870-1911* (1969) and *The Chinese Economy 1912-1949* (1968); Murphey R, *The Treaty Ports and China's Modernization: What went wrong?* (1970); Rudolph and Rudolph, *The Modernity of Tradition* (1969); Levenson J, *Confucian China and Its Modern Fate: The Problem of Intellectual Continuity* (1958); Gadgil D.R. *The Industrial Evolution of India* (1967).

For comparative articles on India and China after Independence, the following in the *American Economic Review* will suffice:

Richman B, *Chinese And Indian Development: An Interdisciplinary Environmental Analysis.* May, 1975; Weisskopf T, *China and India: Contrasting Experiences in Economic Development.* May, 1975. Vol. 65. No. 2; Weisskopf T, *China and India: A Comparative Survey of Performance in Economic Development. Economic and Political Weekly,* February, 1975.

1. Feuerwerker A, *op. cit.*, 1912-1949. p 17.
2. Murphey R, *op. cit.* p 66.
3. Feuerwerker A, *op. cit.*, 1912-1949. p 10.
4. Feuerwerker A, *op. cit.*, 1912-1949. p 6.
5. Maddisio A, *Class Structure and Economic Growth.* pp 61-62.
6. Fairbank & Teng, *China's Response to the West: A Documentary Survey 1839-1923* (1971) pp 53-55.
7. Fairbank and Teng, *op. cit.*, p 62.
8. Fairbank and Teng, *op. cit.*, p 86.
9. Fairbank and Teng, *op. cit.*, p 111.
10. For the introduction of Western technology into India in the eighteenth and nineteenth centuries, see Bernstein H, *Steamboats on the Ganges: An Exploration in the History of India's Modernization Through Science and Technology* (1960).
11. Kidron M, *Foreign Investments in India* (1965). p 14.
12. Fairbank and Teng, *op. cit.*, p 275.

13. Murphey R, *op. cit.*, pp 24-25.
14. Rudolph and Rudolph, *op. cit.*, pp 165-166.
15. Rudolph and Rudolph, *op. cit.*, p 167.
16. Nehru J, *The Discovery of India* (1946) pp 361-362.
17. Erikson E, *Gandhi's Truth* (1970) p 447.
18. Erikson E, *op. cit.*, p 448.
19. Rudolph and Rudolph, *op. cit.*, p 114.
20. For a good description of the "Gentleman-Ideal", see Ortega y Gasset, *Man the Technician,* in Mitcham and Mackay (eds), *Philosophy and Technology* (1974).
21. Mamdani M, *The Myth of Population Control.* pp 56-57.
22. Mamdani M, *op. cit.*, p 58.
23. Mamdani M, *op. cit.*, p 64.
24. For a comprehensive but brief introduction to various aspects of Gandhian thinking on society, education and industrialization, see my, *Interaction of Technologies in India* (1975) Technische Hogeschool, Eindhoven.
25. Darbar G, *Portrait of a President* (1974).
26. I am indebted here to Alitto G.S., *Rural Reconstruction during the Nanking Decade: Confucian Collectivism in Shantung. China Quarterly.* 66. June 1976.
27. Wertheim W.F., *Evolution and Revolution.* pp 137-144.
28. Gopal R, *op. cit.*, p 75.
29. For more detailed information on this issue, see Kidron M, *Foreign Enterprise. op. cit.*, pp 22-23.
30. Kidron M, *op. cit.*, p 24.
31. For an excellent discussion of these issues, see the Mudaliar Lecture of Ward Morehouse, *Science, Technology and National Development in India: The Uses, Non-Uses and Misuses of Research,* published in *Altech,* Journal of the Alagappa Chettiar College of Technology, Guindy. Vol XXI (1972).
32. See Morehouse and Aurora, *The Dilemma of Technological Choice in India: The Case of the Small Tractor in Minerva* Vol XII. No. 4, October, 1974.
33. Kidron M, *op. cit.*
34. Gurley J.G., *Capitalist and Maoist Economic Development,* in the *Bulletin of Concerned Asian Scholars,* Vol 2. No. 3. p 44. See also, Meisner M, *Utopian Goals and Ascetic Values in Chinese Ideology* in the *Journal of Asian Studies.* Vol XXVIII. No. 1. 1968.
35. Quoted in Curwen C, *The Imperial Pushers. TLS.* 11 June. 1976. p 708.
36. Leys Simon, *Chinese Shadows* in *The New York Review of Books.* 26 May 1977.
37. Gittings J, *A Chinese View of China.* p vii.
38. Murphey R, *op. cit.*, p 71.
39. Gouldner A.W., *Marxism and Mao. Partisan Review* (1973).
40. Gray J, *The Economics of Maoism* in Bernstein H (ed), *Underdevelopment and Development* (1973).

41. Mao Tse-tung, *On the Ten Great Relationships* in Schram S (ed), *Mao Tse-tung Unrehearsed* (1974).
42. Bettelheim C, *Cultural Revolution and Industrial Organization in China: Changes in Management and the Division of Labour* (1974).
43. Weisskopf T, *China and India. op. cit.*, p 185.
44. Bettelheim C, *op. cit.*, p 17.
45. Tanzer M, *The Political Economy of International Oil and the Underdeveloped Countries* (1969).
46. Richman B.M., *Industrial Society in Communist China* (1969) pp 640 and 643.

CHAPTER SEVEN

1. Watson A.W., Quoted in Fornaro R.J., *Ideology an·ˈ Archaeolcgy in China, Ecopol* (15 May, 1976) p 744.
2. Leys S, *Chinese Shadows* in the *New York Review of Books*, 26 May, 1977.
3. Hobsbawm E.J., *Labouring Men: Studies in the History of Labour* (1964/1968).
4. Quoted in Heilbroner R, *The Future as History* (1960).
5. Geertz C. *Agricultural Involution: The Processes of Ecological Change in Indonesia* (1963).
6. North D.C. and Thomas R.P., *The Rise of the Western World* (1973). In the university education of my thesis, I have upset the notion that Western medicine was responsible for the rise of the Indian population. Recent support for the ideas I mentioned there has come in from Illich I, *Medical Nemesis* (1975) and from McKeown, T, *The Modern Rise of Population* (1976).
7. Wertheim W.F., *Evolution and Revolution* (1974).
8. Harding, Kaplan, Sahlins and Service, *Evolution and Culture* (1960/1968).
9. Sahlins and Service, *op. cit.*, p 97.
10. Wertheim W.F., *op. cit.*, p 78.
11. Wertheim W.F., *op. cit.*, pp 39, 40 and 41.
12. Heilbroner R, *An Inquiry into the Human Prospect* (1974) pp 94 and 136.
13. Heilbroner R, *op. cit.*, p 89.
14. Heilbroner R, *Growth and Survival in Dialogue* (1973) Vol 6. No. 1. pp 10-11
15. Epstein J, *Capitalism and Socialism: Declining Returns* in *The New York Review of Books*, February 1977. p 37.
16. See Wertheim W.F., *Dawning of An Asian Dream*, Publicatie nr. 20. Afdeling Zuid- en Zuidoost Azie, Anthropologisch-Sociologisch Centrum van de Universiteit van Amsterdam. (1973) 1976. This volume contains a wide range of his articles over the years.
17. Makhijani A, *Energy Policy for the Rural Third World*, (1976) p 2. See also the writings of Lewis Mumford, *The Myth of the Machine* in two parts. For a critique of these views, including those of Rene Dubos;

Jacques Ellul and Theodore Roszak, see Florman S.C., *The Existential Pleasures of Engineering* (1976).

18. Lefeber L, *On the Paradigm for Economic Development* in *World Development,* Vol 2. January 1974. p 1.
19. Kothari R, *State-building in the Third World: Alternative Strategies* in *Economic and Political Weekly* (1972).
20. For a brief introduction to the allopathic system, see Weil A, *The Natural Mind* (1972).
21. For an excellent survey of the distinctive philosophical anthropologies of the civilizations of China, India and the West, see, Marcus J, *Phenomenologies of the Self, East and West* in the *International Philosophical Quarterly* (1972).
22. Rostow W, *The Stages of Economic Growth* (1960) p 12.
23. Woodruff & Woodruff, *Economic Growth: Myth and Reality,* in *Technology and Culture.*
24. Bagchi A.K., *Some International Foundations of Capitalist Growth and Underdevelopment* in the *Economic and Political Weekly,* Special Number. August 1972.
25. For brief introductions, see *Cocoa* and *Bananas,* pamphlets published by the Haslemere Group of London.
26. Most of the quotationary material comes from Barraclough's articles in *The New York Review of Books* on the new international economic order (1975-1976).
27. Lewis A.W., *Some Aspects of Economic Development* (1969) p 15.
28. Lewis A.W., *op. cit.,* p 15.
29. Diaz-Alejandro C.F., *Unshackled or Unhinged? On Delinking North and South.* Paper prepared for the 1980s project of the Council on Foreign Relations, Inc. pp 8-9.
30. Shonfield A, *Trade Within the Undeveloped World* gives an instance of IMF intervention against such trade. The essay may be found extracted in Hensman, *From Gandhi to Guevara* (1969).
31. Lewis A.W., *op. cit.,* p 15.
32. Paige J.M., *Agrarian Revolution: Social Movements and Export Agriculture in the Underdeveloped World* (1975) for a good discussion of the nature of 70 export economies, with some special case studies. See also the illuminating study of Rijpma S, *Basic Agrarian Technology* (unpublished document, University of Wageningen, Netherlands) 1975. For a particular case, see Lower, Dalton, Harwitz and Walters, *An Economic Survey of Liberia* (1966).
33. Bertholet C.J.L., *The Struggle for an Adequate Development Model for the Third World* (1975). pp 48-49.
34. Barnet and Müller, *Global Reach: The Power of the Multinational Corporations* (1974).
35. Perkins D.H., *China's Modern Economy in Historical Perspective* (1975).
36. Weisskopf T, *China and India: Contrasting Experiences in Economic Development* in the *American Economic Review* Vol. 65, No. 2. discusses this conclusion.

37. Müller J, *op. cit.*, claims that the Tanzanian government has wiped out traditional iron-workers more efficiently than the British could have ever done.
38. I am indebted to the essay written by Rawski T.G., *Problems of Technology Transfer in Chinese Industry,* in the *American Economic Review,* Vol. 65. No. 2, for much of the information given here. A more comprehensive essay by the same writer may be seen in D.H. Perkins, *op. cit.*
39. Sigurdson J, *Small-Scale Plants in Cement Industry: Use of vertical shaft Kiln Technology in China* in the *Economic and Political Weekly,* February 1976. p 245.
40. *Science Today* (Bombay). See the special section on cement in India in the January 1975 issue.
41. Kidron M, *Foreign Investments, op. cit.,* pp 312-313.
42. For detailed examination of the exploitative mechanisms involved in the transfer of technology, see Barnet and Muller, who summarize most data, including those of the pioneer, Constantine Vaitsos: *op. cit.*
43. The numerous papers of Ward Morehouse will be helpful to the reader who wishes to study the role of science and technology in India's attempts to meet post-Independence problems. Besides the ones already mentioned, the following need to be carefully studied: Morehouse W, *The Endless Quest: Science and Technology for Human Betterment in India,* Research Policy Program, Lund University, Sweden (1976); *Professional Estates as Political Actors: The Case of the Indian Scientific Community* in *Philosophy and Social Action,* II (4) 1976; *R and D and India's Industrial Future* in the *Economic and Political Weekly,* Vol. V, No. 48 (1970); *The King as Philosopher* in Hoelscher and Hawk (eds), *Industrialization and Development* (1969); and *Scuttling the Coal Gasification Project at Hyderabad: The Perils of Public Patronage in Bridging the Development Gap* (1976) forthcoming in a volume of essays in honour of A. Rahman.
44. Quoted in Onyemelukwe C.G., *Economic Underdevelopment: An Inside View* (1974) p 30. As an African engineer's approach to the problem of Africa's technological development, this volume is impressive and indispensable.
45. See Dias-Alejandro C.F., *op. cit.,* p 16 f.
46. See Kidron M, *op. cit.,* pp 214-215.
47. See the *Times Literary Supplement* (1976). p 952.
48 See Wertheim W.F., *Elite or Vanguard of the Masses? The Chinese Experiment* for a distillation of the views concerning the role of elites in industrial societies.
49. Illich Ivan, *Medical Nemesis* (1975); Mumford L, *The Myth of the Machine* (1967/1970); Schumacher E.F., *Small is Beautiful* (1972).
50. Why should, for example, Mause L (ed), *The History of Childhood* be so called, when every contributor admits that his focus is childhood in the West?
51. Horowitz I.L., *The Rise and Fall of Project Camelot;* Schenk-

Sandbergen L.Ch, *Social Science Research for the American Government in South-east Asia* (in Dutch) in *Wetenschap en Samenleving* (1971) Vol. 25. pp 107-108; Wertheim W.F., *Counter-insurgency research at the turn of the century – Snouck Hurgronje and the Acheh War* in Wertheim W.F., *Dawning. op. cit.*, pp 136-148.

For proposals concerning a "sociology of emancipation", see besides the work of Wertheim, Huizer G, *Applied Social Science and Social Action: A Note on New Approaches* to be included in Huizer and Mannheim (eds), *The Politics of Anthropology; Action Research and Peasant Resistance,* Third World Center, Nijmegen.

52. Quoted in Copleston F, *A History of Philosophy* (1964) Vol 5. Part II. p 120.
53. "But it is of the greatest importance that there is a country – the most huge and most populous one in the world, a continent in itself – where exists, at least, an awareness of such problems, and where there is a constant struggle to cope with these in a cautious and deliberate, but bold way." Wertheim in *Polarity and Equality in China* (with L.Ch. Schenk-Sandbergen) 1976. University of Amsterdam.
54. Diaz-Alejandro, *op. cit.*, p 50.
55. The ideology round which the World Development Movement in England bases itself; see also the *Haslemere Declaration.*
56. The Dutch economist, Jan Tinbergen, has also asked multinationals to work in this direction. See his RIO report (1976).

    The thesis that the extension of the Western world's technology to the Southern countries has created at least widespread ecological damage, may be found argued in Farvar M.T. and Milton J.P., *The Careless Technology: Ecology and International Development* (1972).
57. Barnet and Müller, *op. cit.*, p 151.
58. "As a rule, innovations introduced by members of an urban or urbanized elite do not contribute to modernization but rather to impoverishment of the peasant masses." Wertheim in *Polarity and Equality in China, op. cit.*, p 42. But this is also the principal thesis of the dependencia school and Andre Gunder Frank's "development of underdevelopment". See also E. Vallianatos, *Fear in the Countryside: The Control of Agricultural Resources in the Poor Countries by Non-Peasant Elites* (1976).
59. George S, *How the Other Half Dies* (1976). The one "good" example Ms George gives of India is a phony.
60. For a good brief account of the Green Revolution, see the report of the North London Haslemere Group, *The Death of the Green Revolution;* for the role of multinational corporations and their interest in the Green Revolution, see the work of E. Feder, *The Economic, Social and Political Impact of Agricultural Modernization on the Latifundio Agricultures of Latin America* (Epitome) (1974); *The New Penetration of the Agricultures of the Underdeveloped Countries by the Industrial Nations and their Multinational Concerns,* The Hague. Feder has also termed the new business in agriculture in the Southern nations, "McNamara's Little Revolution".

61. For a great deal of inappropriate technology, see Farvar and Milton. *op. cit.*
62. Van Bronckhorst B, *Development Problems in the Perspective of Technology* (1974) p 13.
63. The fact that the seven sisters controlling the energy industry have sabotaged the solar energy programme is, I suppose, quite common knowledge by now.

Index